The Making of th

Ulrich Pfammatter

The Making of the Modern Architect and Engineer

The Origins and Development of a Scientific and Industrially Oriented Education

Birkhäuser – Publishers for Architecture
Basel • Boston • Berlin

This publication has been generously supported by:

Gerold und Niklaus Schnitter Fonds zur Förderung der Technik-geschichte at the ETH;
The foundation for promoting the University of Applied Sciences Lucerne,

as well as by the following Swiss industrial firms:

Eternit AG Niederurnen
Holderbank Financière Glarus AG
Jansen AG Oberriet/SG
Otis AG Fribourg
Schmidlin AG Aesch/Basel
Sulzer Infra Lab AG Winterthur
Glas-Trösch Holding AG Bützberg
USM U. Schärer Söhne AG Münsingen

Translation from German into English: Madeline Ferretti-Theilig

A CIP catalogue record for this book is available from the Library of Congress, Washington D.C., USA.

Deutsche Bibliothek Cataloging-in-Publication Data

Pfammatter, Ulrich:
The making of the modern architect and engineer / Ulrich Pfammatter. [Transl. into Engl.: Madeline Ferretti-Theilig]. - Basel ; Boston ; Berlin : Birkhäuser, 2000
Dt. Ausg. u.d.T: Pfammatter, Ulrich: Die Erfindung des modernen Architekten
ISBN 3-7643-6217-0

This work is subject to copyright. All rights are reserved, whether the whole or part of the material is concerned, specifically the rights of translation, reprinting, re-use of illustrations, recitation, broadcasting, reproduction on microfilms or in other ways, and storage in data bases.
For any kind of use permission of the copyright owner must be obtained.

This book is also available in a German language edition (ISBN 3-7643-5473-9).

© 2000 Birkhäuser – Publishers for Architecture, P.O. Box 133, CH-4010 Basel, Switzerland.
Printed on acid-free paper produced of chlorine-free pulp. TCF ∞
Printed in Germany
ISBN 3-7643-6217-0
ISBN 0-8176-6217-0

9 8 7 6 5 4 3 2 1

Contents

Preface

Allow me to contribute a few comments on the present monograph in its relation to the author, as it cannot be his responsibility to introduce himself.

The École Polytechnique (founded in 1795) and the École Centrale des Arts et Manufactures in Paris were the starting point and central topic of Ulrich Pfammatter's scholarly research, which found expression in a doctoral thesis. However, it would be false to endeavor to class that work as well as the present monograph "The Making of the Modern Architect and Engineer. Origins and development of a scientific and industrially oriented education" as a commemorative volume on that school's bicentennial. On the contrary, this book fundamentally explores and questions without adhering to a simple historical and documentary perspective. How did the modern profession of the architect evolve? What can be learned from the changes it underwent?

Pfammatter was educated as an architect at the ETH in Zurich; through studying education sciences at the University of Zurich he became acquainted with modern European didactic models. His training under Bernhard Hoesli and working with Heinz Ronner allowed him to become familiar with the methodology of architectural engineering history and, most importantly, with the concept of "Basic Design" (Zurich Model). Questions regarding the foundations and basic principles of architects originate from this. Pfammatter's own teaching work in the subjects of design, construction and the history of architectural engineering added depth to this very unique and synoptic perspective of his on the present theme.

Issues that usually belong to and remain in the sphere of the historian are approached here from a new and different perspective. Through analyzing the conditions of the time, determining the diverse contingencies as well as by continuous reflection upon the didactic models, methods and procedures the present monograph is endowed with an operative quality. By unraveling new sections of the history of inventions, by questioning the achievements of educational institutions as "schools of thought", by searching out the very origins of today's architectural and engineering profession this work makes a major contribution. The author thus augments our understanding of history, which according to Claude Lévi-Strauss is of great importance to every generation in order to be able to assume responsibility for one's own future.

Architects and teachers of architecture are today once again more than challenged to attempt a re-orientation based on basic and also historically established insights into the nature of their profession. I believe that this new beginning, two hundred years after the epoch-making foundation of the École Polytechnique, is of existential importance.

Herbert E. Kramel, ETH Zurich

Introduction – The Modern Architect and Engineer: The Development of a New Tradition

Setting up science as a practical science occurred during the era of the Enlightenment by philosophers and scholars who not only wrote treatises and worked in scientific academies but also took part in technical and proto-industrial projects that were part of and foreshadowed a practical re-shaping of the society and state. In the second half of the 18th century "Philosophes" such as d'Alembert, Turgot and Condorcet – in America, Franklin and Jefferson, in Germany, von Humboldt – considered the development of the sciences and technical arts to be the motor for improving the general welfare and individual happiness of free and equal citizens within the framework of a democratically organized state, which was to be supported by public education.[1] First in France, followed by Austria and America and later also Germany, the tradition of polytechnical and industrial education of architects and engineers was established. [color plate 1]

The invention of a systematic and universal teaching model that was to serve the general public, its first institutionalization at the Parisian École Polytechnique in 1794/95 and its further, industrial development at the École Centrale des Arts et Manufactures, founded in 1829 equally in Paris, brought about *the modern architect*. While the modern engineer of bridge and road construction had already been "invented" at the Parisian École des Ponts et Chaussées[2], which was founded in 1747, the establishment of a tradition concerning the modern architect is the feat of the first veritable teacher of architecture at the École Polytechnique: Jean-Nicolas-Louis Durand, who as of 1797 educated many generations of architects and engineers (of building construction) as well as of teachers over a period of thirty years.

Three pillars of the modern architect. Due to historical circumstances (the re-formation of post-revolutionary society, belated French industrialization in comparison to England) first of all a *new understanding of praxis* evolved as a reaction to the altered responsibilities of the young Republic. Thus, not only

[1] Hager, 1986.

[2] Picon, 1992; the title of this present study refers to this work published by A. Picon.

[3] These include small to large factories, housing and living quarters with heating and sanitary facilities, hospitals and schools, market halls and warehouses, cultural buildings, also infrastructure facilities for canal shipping and harbor complexes, later for railroad engineering and not least for city utilities (canalization systems, traffic facilities and electrification) and much more.

[4] The modern concept of teaching replaced the classical learning

were political edifices of representation, revolutionary monuments and structures for republican celebrations primarily in demand but increasingly buildings for government institutions and the new societal life.[3]

Secondly, the pressing industrialization required up-to-date training and a *novel model of teaching.* The changing demands in praxis necessitated new schools and a transition from the traditional master studio at the academies to an institutionalized curriculum.[4] The teaching and learning models invented at that time comprised, next to practice-oriented lectures, exercises and projects in the drawing classes, practical courses in the laboratory and studio, excursions, construction site visits, vacation and field work as well as training, etc. In addition, academic guidance in the form of a permanent teaching staff with tutors and assistants as well as regular examinations, internal class competitions and graduation diplomas can also look back on a two-hundred-year history. This teaching model was characterized by a form of "encouragement pedagogy" that made great demands upon the teacher to "fire the

model of the master's studio, however without giving up the advantages of, for instance "learning by role model", precision in work techniques and so on. In this respect it must be noted that a branch of educating architects occurred over a long period of time at the classical academies such as the École des Beaux-Arts , which still do so today. On the role of the fine arts system in the USA, cf. chapter on "American Lines of Tradition".

Scientific collection of the Académie des Sciences in the Louvre (18th Century).
(N. Dhombres, 1989, p.24)

students' ambitions and learning motivations and to never cease in endeavoring to have their work achieve the greatest possible perfection."[5] This basic educational attitude – which encouraged the students' interest in their work, their identification with the task at hand as well as a sense of responsibility toward their own work while also endeavoring to organize the class within an atmosphere of competition so that the student's achievements could be perfected – originates in the philosophy of the Enlightenment which considered mankind to be inexhaustible in his desire for knowledge and in his ability to learn, and basically saw in all men the same predisposition to learn as well as the same drive for "perfectibilité"[6].

[5] Monge, 1794, p. 2.

[6] Moreover, it corresponded with a postulate of the Enlightenment to provide all men, independent of their background and social status, the ability to acquire education; the institutionalization of a standardized admission procedure as well as a extensive system of stipends for all interested and well-prepared candidates belongs to the tradition of these modern schools of higher scientific and technical education from the beginning.

Thirdly, the new understanding of theory and praxis resulted in an industrially oriented curriculum. *Modern architectural theory* after 1797, as a consequence of erratic governmental, social and industrial development, thus initially established the equivalence of "construction" and "composition" and later the primary importance of construction as an architectural dimension of design. In addition, the paradigm shift in architectural theory followed the demands made by industrial culture and technology for precise designs as a reaction to the functional nature of the demand. The doctrine of "usefulness" (utility value) that connects Durand with those of the "economy of construction" formed the basis for composition of which the highest quality was still being demanded. In this respect the "school of style" was overcome and evolved into a "school of method".

While in Durand's architectural teaching an expanded "Cartesian process of knowledge" was transported by linking a deductive teaching method with an inductive learning process, Charles-Louis Mary, the leading teacher of architecture at the École Centrale, invented a form of "layered model" of learning: a methodical correlation between theoretical basic subjects, applied exercises and practical courses as a continual system with increasing weight placed on reference to praxis and on professionalism in the working method. The transfer between theory and praxis was therefore carried out for the first time at this school.[7] In its teaching program it also replaced mathematics, a

[7] However, at the same time, Durand's principle of learning from "building types" was retained and the tenets of "building types" were maintained unbroken for over 150 years in architecture schools. The further shift in paradigm that took place in architectural teaching, which arose from the transition to learning from "problem types", did not occur until 1960 at the Federal Institute of Technology in Zurich; cf. the chapter on the future outlook at the end of this book.

science which penetrated the entire curriculum at the École Polytechnique, and increased the importance of chemistry, which paved the way for a modern instruction in building materials with iron, steel, glass and concrete. The kind of contribution to the development made by successive schools in Europe (Karlsruhe, Zurich, etc.) and America occurred on the level of school and study organization, of the curriculum and at the most on the level of project tasks. However, it was these two first French schools that founded the modern teaching model still in effect today.

Schools of thought in the history of architectural engineering. The learning institutions for architects and engineers treated here belong to the "schools of thought" that advanced scientific, technological and industrial progress, and also especially contributed to the history of architectural engineering. The methodical and didactic innovations that came into effect in the context of modern teaching programs and curricula influenced major inventions of architectural engineering. Solving the problem of theory and praxis in the teaching models of the polytechnical-industrial tradition acted as an "educational motor" for the history of technology.

The teaching pioneers – at the same time leading individuals of science, technology and industry who the schools engaged as professors – belonged to teaching groups or scholarly circles that exerted a major influence on the school culture and created centers of scientific discourse. In the analysis of each school biography it becomes apparent that, through its professorial "corps" and the quality and kind of praxis-oriented instruction given, only in specific phases of its history was every school able take on a leading role and exert a major influence on students who then again were active as pioneers or groups in science, technology and industry or in practice, research and teaching, thus strengthening the effectiveness and reputation of their "alma mater".[8]

From this perspective the teaching and learning methods used, which have been very little considered in literature, are granted a re-assessment here with regard to their contribution to the history of architectural engineering. Therefore, the question is also di-

[8] In the transition to a standardized types of training, which was characteristic of both the École Polytechnique and the École Centrale, to a system of departments or profession-specific courses of study (Karlsruhe, Zurich and others) each department must additionally be considered in their independent development in order to discern periods of "flourishing" of particular school environments.

rected towards the educational background of those pioneers of building history: who were the students, who were their teachers? What and how did they learn and teach? Architects and engineers were and still are not only witnesses of epochal processes of development but also protagonists in that process. The same applies to the schools.

"Vacation projects" at the École Centrale. A form of "vacation or field work" instruction introduced by the École Centrale around 1836 illustrates the theory-praxis-connection. It entailed a comprehensive analysis of a complex process of manufacturing. Through the high qualitative and quantitative demands made upon the students they were able to acquire a precise picture of the functional relationships between man and the machine, of manufacturing facilities and processes, of how work is organized and companies are managed in a way offered by no other school of that time. Human productivity, the share of mental and physical work, the state of material and physical well-being as well as the security measures installed at the workplace were given just as much notice by the students as the finished products and their use. Graduates of this school were not only to fulfill the most different functions in science, technology, industry and society but were also to become conscious of their social responsibility. A key element in this process was to surmount the dichotomy between the occupation of architect and engineer and as a consequence to take advantage of the new perspective of interdisciplinary thought, work and action. Behind this programmatic-didactic task lay a philosophic approach, which not only carries the mark of the intellectual culture of the "schools of thought" of the time – for instance Saint-Simon's positivism – but also opened perspectives on an evolutionary, non-liberalistic-permissive attitude towards scientific technological progress.

An "engineer-architect" educated at modern schools was therefore not only "modern" in his thought and actions because he took on novel building tasks or worked and experimented with iron, glass and concrete. He was modern because he considered his discipline to be part of a whole, as a means to create structures of culture and social utility. In this respect the era in which a new – material as also mental –

tradition of building and education became manifest acted as a challenge of cooperation between architect and engineer and as a chance to bring forth novel achievements.[9] Since the industrial revolution every challenge that has been met is to be understood along this line. Not only does a thoughtlessly brought about crises concern the architect and engineer, science and technology, but also the responsible resolution of that crises – the protagonists are all the same in the end.

[9] Giedion also points this out: "The Division of Tasks between the Architect and the Engineer" in: Giedion, 1941, p. 146–152.

On the method. The first industrial revolution as well as the philosophy of the Enlightenment and of the French and American Revolution – which brought about the schools, teachers and students presented here – not only meant that a change in system occurred in the history of science and technology, but it also represented a point of transition in the history of ideas. This, then, also influences the historical perspective applied to that process. Picon refers to this when he speaks of "schools of thought" that had a major influence on technological revolutions. The technological *systems* at the historical point of transition to the industrial revolution (end of the 18th to middle of the 19th century) were according to Picon not only influenced by organizational and professional *structures* in the context of scientific and technological development but were superimposed by the intellectual activity of pioneers, scholars, engineers, technicians, constructors, even skilled workers, who represented the *patterns of thought* of the era. These, again, became personally manifest or became "material reality" in the context of project groups, scientific associations, or professional unions.[10]

[10] Picon, 1996, p. 39.

This approach will be applied in the present study on "schools of technological thought" of epoch-making educational institutions for architects and engineers. At the same time it was necessary to also focus attention on the uniquely individual characteristics exhibited by the protagonists in the process of transition: *the persons* developed into educational, scientific and technological innovators in the context of professional milieus, thus forming their "physiognomy". By using case studies it will be analyzed and revealed how schools reacted to scientific and technological developments, which teachers were engaged and in what kind of intellectual and practical milieu young students were educated to be pioneers of the modern world, and which individual characteristics exerted an additional influence in that process.[11]

The presentation of these insights is motivated by the objective to understand the activity of those promoters and groups within the context of their era. Correlations are often made between biographies and school histories and their historical and cultural background, as the available literature often refrains from mentioning in which schools, under which teachers and in which educational context architects and engineers were educated.

The present work was based on the author's dissertation[12], which has been condensed as well as expanded in order to place more emphasis on the correlations between the school histories and the history of architectural engineering as well as on relevant international developments. Mostly primary sources have been used as well as school chronicles, student lists, diploma projects, text books, treatises etc., also individual reports that are able to authenticate a close connection to the persons, groups and events treated or guarantee contact with the following generation. The philosophical section of the dissertation has been left out here and has been summarized in core statements to be found in separate and marked text blocks.

Limitations. The object was not to compile a systematic study of all countries. The period of time taken into consideration has been limited within a further and more narrow context: within the period of the Enlightenment and industrial revolution up to the threshold of modern architecture on the one hand, and on the other as a core period the era between the founding of the École Polytechnique in Paris (1794/95) and the conclusion of the consolidation phase of the Swiss Institute of Technology in Zurich (around 1870), two dates which represent historical turning-points in Europe as well as in America. However, this work does not refrain from addressing current issues when appropriate.

It is necessary to delineate clearly from the contextual implications of the term "polytechnical education" as used in Marxist ideology primarily after the Second

[11] Moreover, the school-historical, curricular as well as intellectual environment never exerts just – one dimensional or mechanistic – influence neither on the thoughts and deeds of a school graduates if he develops his practical work, nor on his solution, procedure strategies or his innovations, etc.; Ferguson shows additional implications of factors influenced by personality: Ferguson, 1992, especially chapter 7.

[12] Pfammatter, 1995.

[13] Although Marx himself did not use this term, he is mentioned in academic literature, on the one hand in the context of the

World War.[13] In this context the Marxist concept of "combining instruction with productive work" is geared toward instrumentalizing the idea of education for the ideological as well as productive achievements demanded as part of its government and party program. This idea is to be found neither in the educational ideas nor in the concept of mankind held by those philosophers of the Enlightenment who were influential in the formation of the polytechnical model of teaching as analyzed in this work. On the contrary, it is to be found in materialistic-mechanistic conceptions[14] that generalize and reify "man's nature", thus denying and combating the idea of individual value, individuality and the freedom of man as person, as was exemplified especially by the Jacobin rule of terror immediately before the founding of the École Polytechnique.

Acknowledgments. I am deeply thankful to both examiners of my dissertation, Prof. H.E. Kramel from the Department of Architecture at the ETH in Zurich and Prof. Dr. F.-P. Hager (†) from the University of Zurich (Historic-Systematic Pedagogy). While Kramel encouraged me to begin with the first institution offering a systematic and curricular education in architecture, hence with Durand – and not already with Imhotep, the Egyptian High Priest architect who lived 4,500 years ago – my attention to philosophy and pedagogy was directed by Hager towards the origins of modern institutions of education within the system of thought of the Enlightenment. A different approach to the theme of architectural history and its protagonists as well as to principles of teaching and methodology in education were provided by my former teachers at the ETH Zurich, Bernhard Hoesli, Heinz Ronner, Herbert E. Kramel and Adolf Max Vogt.

Moreover, I am indebted to the team of librarians in Paris who supported my work: Mme N. Magnoux at the École Centrale Paris, Mme F. Masson and Mme Brenot (École Polytechnique), the archive of the Muséum d'histoire naturelle as well as Mme F. Harmon, who copied many of the "Centraux'" final projects.[15] In Zurich the ETH main library, the architecture library and the collections of science history (Dr. B. Glaus) assisted in making the sources needed available. Finally, I would like to thank all my colleagues and friends who

"socialist school" of the GDR in the 1950s, and on the other hand in the educational policy publication "Neue Linke", which in the aftermath of the student revolution of the 1960s adopted this model and modified it for Western European-American use, cf. for instance Krapp, 1971.

[14] Holbach's "Système de la Nature" or La Mettrie's "L'Homme Machine" can be considered as French revolutionary precursors of this trend, a materialistic approach which results in a new form of Determinism which as a dogmatic concept was contrary to the free spirit of the Enlightenment and which represents an isolated phenomenon rejected by the Encyclopedists Voltaire, d'Alembert and Condorcet (on the occasion of their meeting at Voltaire's estate in Ferney near Geneva in the year 1770). The philosopher Ernst Cassirer states the following: "The convictions of the encylopedian group are not represented by Holbach and La Mettrie but are represented by d'Alembert." Cf. Cassirer, 1973, p. 73 and Badinter, 1989, p. 68 f.

[15] Authorized by M. Bernard Toulier, Conservateur de l'Inventaire Général, who I would also like to thank here for granting permission to reproduce (for illustrations marked by ©)

provided me with their insights and materials as well as my parents who encouraged me as an architect and citizen to become interested in the times and in history.

The present book was generously supported by the Gerold und Niklaus Schnitter-Fonds zur Förderung der Technikgeschichte at the ETH, the foundation for promoting the University of Applied Sciences Lucerne, as well as by the following Swiss industrial firms: Eternit AG Niederurnen, "Holderbank" Financière Glarus AG, Jansen AG Oberriet/SG, Otis AG Fribourg, Schmidlin AG Aesch/Basel, Sulzer Infra Lab AG Winterthur, Glas-Trösch Holding AG Bützberg, USM U. Schärer Söhne AG Münsingen. This financial assistance permitted the book to have a high-quality design and layout.

My thanks and appreciation go to Madeline Ferretti-Theilig for the difficult, expert translation demanding great sensitivity. I would like to express my deepest gratitude, however, to my wife Johanna, who followed my work with lively interest and in our discussions developed questions that resulted in ever new reflections, reconsiderations and corrections.

Islisberg, 1999 *Ulrich Pfammatter*

Sources and References to Further Reading

Badinter, E. and R., Condorcet. Un intellectuel en politique, Paris 1989.

Cassirer, E., Die Philosophie der Aufklärung, Tübingen 1973 (French edition: La Philosophie des Lumières, Paris 1990).

Dhombres, N., Les Savants en Révolution 1789–1799, Paris 1989.

Ferguson, E.S., Engineering and the Mind's Eye, The MIT Press, Cambridge, Mass., 1992.

Giedion, S., Space, Time and Architecture. The Growth of a New Tradition, Cambridge, Mass., 1941.

Hager, F.-P., Zum Begriff der Aufklärung in der englischen, französischen und deutschen Aufklärungsphilosophie (1986), in: id. (Ed.), Wesen, Freiheit und Bildung des Menschen, Studien zur Geschichte der Pädagogik und Philosophie der Erziehung, Vol. 10, Bern 1989, p. 227–243.

Krapp, G., Marx & Engels über die Verbindung des Unterrichts mit produktiver Arbeit und die polytechnische Bildung, Frankfurt a/M. 1971.

Monge, G., Stéréotomie, in: Journal Polytechnique, ou Bulletin du Travail fait à l'École Centrale des Travaux publics, Paris 1794 [École Polytechnique, Archives: J.E.P., Tome I, Cahier 1, 1794].

Pfammatter, U., Ursprung, Entwicklung und Bedeutung des polytechnischen und industriellen Unterrichtsmodells. Elemente zur modernen Architekten- und Ingenieurausbildung, Diss. ETH Zurich 1995.

Picon, A., L'invention de l'ingénieur moderne. L'École des Ponts et Chaussées 1747–1851, Paris 1992.

Picon, A., Towards a History of Technological Thought, in: R. Fox (Ed.), Technological Change. Methods and Themes in the History of Technology, Amsterdam 1996, p. 37–49.

The École Polytechnique in Paris

The École Polytechnique, which was founded in the winter of 1794/95, was not only the first training center for engineers and architects with an institutionalized teaching program and with modern training goals. Those scholars active at the École Polytechnique, especially Gaspard Monge and Jean-Nicolas-Louis Durand among others, introduced a research and technically oriented teaching program with a unique curriculum for the master-builders of the approaching industrialized era, and established a novel teaching model, the "Modèle polytechnique" which was imitated in Europe and the USA in the first decades of the 19th century. During the first two decades of its inception the École Polytechnique, as an academic and technical corporate body consisting of its teachers and graduates, established itself as a school of thought, whose methodology exerted a major influence on the history of technology, particularly also on architectural engineering.

The École Polytechnique – A Child of the French Revolution

The establishment of the École Polytechnique was based on the achievements of scholarly groups who, during the "Ancien Régime", distinguished themselves by their pioneer inventions, were active as scientists and professors and were known throughout the world by their publications. During the Revolution they pursued their plan to establish a training-ground for cadres of engineers, architects and officers who were to assist in the creation of the new state by guaranteeing its organization and infrastructure and who were able to serve in a modern army. The initiators included the chemists Berthollet, Chaptal, Fourcroy, Guyton de Morveau, Pelletier and Vauquelin, the physicists Hassenfratz and Barruel, the analysts Lagrange and Prony, the mathematicians Monge and Hachette, the engineers Lamblardie and Prieur, the architects Baltard and Delorme, the fortification engineers Dobenheim and Martin as well as the painter Neveu. This group comprised the founding generation and the first teaching staff of the École Polytechnique[1].

While as of January 1792 (still during the Revolution) a *Comité d'instruction publique* was formed under the chairmanship of the mathematician and philosopher of the Enlightenment Condorcet to develop the basic principles of a new and comprehensive national system of instruction (Rapport et Projet de Décret[2]), a year later a *Commission des travaux publics* was also set up by the constituent assembly (Convention), which under Fourcroy's leadership developed a plan to establish a technical institution of higher learning for engineers and officers based equally on the education ideals formulated by the philosophy of the French Enlightenment.

On the development of the Revolution after 1789. The Ancien Régime under Louis XVI initiated its dissolution with the summoning of the "Estates-General", which had not met since the reign of Louis XIV. On the day of its convention, the 17th of June 1789, this body, which was renewed in preliminary elections, declared itself the "Third Estate" (Tiers État) and represented the modern bourgeoisie, the progressive nobility as well as the people's broad interests. It was concerned with replacing the economy and policy of privilege maintained by Absolutism with a modern government that supported a free economy and democratic rights. June 17, 1789 instituted a legislative revolution and primarily aimed at establishing a constitutional monarchy. Its most important achievements include the Declaration of the Rights of Man (26 August 1789) and the adoption of a constitution (3 September 1791). However, more radical forces sought a more comprehensive revolution and organized the storm on the Bastille. This second, violent revolutionary stage brought about repeated political attacks against the legislative system, which in the autumn of 1793 finally culminated in the reign of terror under the Jacobin faction led by Robespierre ("Terreur"). It was brought to an end on 28 July 1794 and replaced by a Republic, which was supported by a broader political base ("Thermidor"). Yet it could not prevent a new form of despotism from developing as of 1799, which finally reached its apex in General Napoleon Bonaparte's "Empire" (until 1815).[3]

[1] Pinet, 1887, Chapter "Première Organisation", p.377ff., esp. p.383.

[2] Condorcet, 1792.

[3] Michelet, 1952 (1847); Griewank, 1958; Schmitt, 1980; Soboul, 1989; Furet and Ozouf, 1992.

The Role of Scholars During the Revolution

Some of the scholars and engineers mentioned above and engaged in promoting the new School of Polytechnics assumed offices and responsibilities at the outcome of the Revolution within the context of the new institutions. Thus Monge, one of the most famous mathematicians, was engaged as minister for the Navy under Danton (before the Terror), and his former students from the École militaire de Mézières, Carnot and Prieur, became members of the Committee of Public Safety (Comité de salut public), the latter even during the Jacobin rule of terror.

Because the French scientific elite addressed and cooperated on common tasks during the Revolution and prior to the establishment of the École Polytechnique, the core ideas and concepts of the philosophy of the Enlightenment were able to be introduced into individual projects, particularly in the areas of democracy and education, despite the anti-scientific predilections on the part of revolutionary groups. These included, besides the Declaration of the Rights of Man and the republican constitution, the founding program of the École Polytechnique (1 September 1794) directly after the fall of the Jacobin dictatorship and the establishment of other institutions of education such as the École Normale or the Conservatoire des Arts et Métiers in the same period of time. [4]

[4] On the history and importance of the Conservatoire des Arts et Métiers: Le Moël, Saint-Paul (Ed.), 1994.

The union of enlightened scholars during the Revolution and their impact on the future system of education as symbolized by a relief portraying the chemists Guyton de Morveau and Berthollet, the natural scientists Vicq D'Azyr and Lamarck as well as the mathematicians Condorcet, Lagrange, Laplace and Monge. (Nicole Dhombres, 1989, p.15)

The French historian and mathematician Jean Dhombres has termed this group of dedicated scholars as the "Lobby scientifique".[5] A few of its members also worked as "Législateurs scolaires"[6] and were therefore able to exert influence not only plans for future school and higher education systems but also on scientific activity.

[5] Dhombres, 1987, p.19ff.

[6] Julia, 1981.

The continuity of educational ideals as developed in the intellectual atmosphere of the Ancien Régime was programmatically secured by this group of scholars. This corresponds with a particular phenomenon that Alexis de Tocqueville, although with regard to state and constitutional ideals, also described.[7] During the course of the Revolution these principles of the Enlightenment were continuously contested, which is particularly evident in the educational debates that took place in the National Convention over the concepts submitted by Talleyrand through Condorcet up to Robespierre.[8]

[7] De Tocqueville, 1856.

[8] Guillaume, 1948.

"An Appeal to Scholars"

In an Appeal from 11 March 1793 a few of the leading scholars mentioned above, namely the chemists Berthollet, Chaptal, Fourcroy, Guyton de Morveau and Vauquelin, the physicist Hassenfratz, two engineers, Lamblardie and Prieur as well as the mathematician Monge were brought together by the Convention in a commission of public works (commission des travaux publics) under the leadership of Fourcroy and ordered to establish an École centrale des travaux publics, the direct precursor to the École Polytechnique, and to appoint a highly qualified faculty.

However, the events of war forced France to quickly put together a scientifically and technically trained cadre of officers in order to satisfy its most immediate needs with regard to arms and munitions production. Thus, already one year later the task given to scientists' was to produce munitions and weapons (salpêtre, poudre et canons) quickly and efficiently. The first school to be founded after the cataclysmic events of 1789 was therefore not an École Polytechnique but the "École de Mars". As of 1 June 1794 it offered an accelerated course that was to last four months. 3,000 young citizens camped on the "Plaine des Sablons" outside Paris and were trained in military tactics and republican convictions. It was the time of the Jacobin reign of terror, the third phase of the Revolution. Those engineer-officers trained at the École de Mars were to be deployed in various parts of the country as organizers and directors of similar training and recruitment schools. History, however, took a different turn.[9] [color plate 2]

[9] On the École de Mars: Julia, 1981, p.12; the training project was documented by Guyton, 1794, the program of the Comité de salut public, 1794.

The Chronicle of a Development

The École Polytechnique's biography presents very unique characteristics. These are marked by its affiliation to the government and its institutions – such as the military sector, by the cultivation of a corps spirit and finally by a traditional reference to its revolutionary origins. This has resulted in a dependence on political developments which has characterized the school's entire history. The new school was imparted the particular task of contributing to the aggrandizement of France's status in the world, especially with regard to its competition with Great Britain. Founded upon a dynamic relationship between its governmental-regulatory and popular-revolutionary predilections the demands upon the approach to be taken by the "Polytechniciens" were formed over time – from within as from without. The students' active participation in the political upheavals of 1830 and 1848, for instance, correspond with the "École's" revolutionary origins. Still today a red and black uniformed group of Polytechniciens leads the parade march on the Champs-Elysées on the 14th of July!

The Preparations for an École centrale des travaux publics

As immediate successor to the École de Mars and directly after Robespierre's overthrow the plans formulated in the Appeal to Scholars to establish an École centrale des travaux publics, interrupted by the Jacobin dictatorship, were now able to become realized under the new political climate of the Thermidor.

However, first 300 candidates were to be prepared for the new school over a period of three months within the context of a condensed course of study. Each of the candidates chosen by entrance exams held in 22 French cities were able to enter the first, second or third levels according to their educational background so that at the moment the regular courses were opened they could take part in all three year levels at the same time and to the same extent.[10]

[10] On the preparatory courses: Langins, 1987, esp. Chapter I-IV, as well as reprints of the training statutes and curriculum included in the appendix.

In addition, the entire faculty was also occupied in this manner from the very beginning. This is significant because since all higher learning institutions had been closed, particularly however since the time-honored Académie des Sciences was dissolved on 8 August 1792, the entire body of world-famous French scholars were without secure employment and had hovered on the fringes of political priorities (apart from the École de Mars), and were thus also without access to research institutions.

Among the newly founded research institutions immediately established after the overthrow of the Jacobin dictatorship, which took place on 25 October 1795 – the Institut National, École Normale, Conservatoire des Arts et Métiers – the École centrale des travaux publics played a major role, as with it a completely innovative form of school and training was invented and installed. As a consequence, a new direction was also set towards a praxis-oriented, methodical educational environment. Previous major schools for engineers and officers from the era of the Ancien Régime that slowly re-opened after the revolutionary turbulence were augmented by a further institution at the base. It was to provide all interested and well-prepared candidates access to application-oriented specialized schools of higher learning in the civil and military sector by means of a broad, yet standardized and comprehensive course of study.

Reference to the Thought of the Enlightenment

The new school was formed in the spirit of the Enlightenment with the aim to establish the technical arts, industrial production and therefore also the corresponding activity of engineers and architects upon a scientific basis. It was supported by the Enlightenment's ("Lumières") reasoned conviction that science and technology are capable of bringing about human progress, of promoting the welfare of all and thus also the happiness of the individual as a free and equal citizen. Particularly the last of the philosophers of the Enlightenment, Condorcet, following on d'Alembert's ideas, supported the opinion that the knowledge acquired through instruction and education essentially contributes to the correct application of the newly declared Rights of Man and the established constitutional social organizations, and that only this is able to protect democratic achievements from despotism, errors, superstition and an arbitrary state of law.[11]

[11] Cf. Condorcet's statements in his five "mémoires" ("Sur l'instruction publique") on issues of instruction, education and democracy from 1791, in: Oeuvres, Vol. 7, 1847, p.169–448; Hager, 1993.

Following the scientifically reasoned system of education, as already formulated in d'Alembert's preface (Discours préliminaire) to the Encyclopedia[12], one considered school – in contrast to Anglo-Saxon teaching and learning models of apprenticeship or shop culture – to be the central site for imparting knowledge and for acquiring capabilities and skills. The knowledge to be imparted was divided, in accordance with the encyclopedic system, into two major areas: mathematical methods of computation and physical principles. Both disciplines together formed the core of the curriculum and their interconnection, theoretical as well practical, was to become a characteristic of polytechnical education. The school of the post-Enlightenment, however, is not only the site of reflection and research of abstract knowledge but also of the energetic application of reasoned activity to serve the general welfare and of the practical realization of knowledge in the form of useful achievements of civilization: Practice is to be "enlightened" – thus the term Lumières.

[12] D'Alembert, 1751.

Jean-Lerond d'Alembert (1717–1783), mathematician, philosopher of the Enlightenment, patron of the French sciences, co-author of the Encyclopedia (together with Diderot). (Durant, 1935, p.257)

D'Alembert and the Encyclopedia. Acquiring knowledge according to d'Alembert's "Dicours Préliminaire" of the Encyclopedia of 1751 occurs by way of three epistemological categories or "possibilities": by memory and obtaining knowledge (Mémoire), by reason or the power of reasoned thought (Raison), or by the power of imagination (Imagination). D'Alembert placed the field of history, human works and nature in the first category, philosophy, ethics and the natural sciences in the second and the fine as well as applied arts in the third. D'Alembert went on to differentiate each of these areas within a elaborate systematic outlined in an encyclopedic chart. The "Encyclopédie", published by Diderot and D'Alembert and with the cooperation of authors such as Voltaire, Condorcet and others, ultimately illustrates this systematic and aims at informing the public of the state of the sciences and arts and of the correlations between them.[13]

[13] D'Alembert, 1751.

Gaspard Monge – Promoter of the École Polytechnique

Gaspard Monge, the actual founder of the new polytechnical school, was a mathematician closely affiliated with the "Philosophes" and with regard to the school's novel model of instruction an inventive interpreter of the Lumières' concepts. The division of subjects and their allocation within a curricular system imbued with mathematics as well as the simultaneous arrangement of the entire canon in coherence with an instructive and illustrative discipline – the "Géometrie descriptive" – which made possible a practical application of the theoretical principles of all then existing fields of engineering, including architecture, was Monge's invention. This will be treated in more detail later on.[14]

[14] On Monge: Dupin, 1819; Taton, 1951.

The Founding Document

The founding document of the new École centrale des travaux publics was formulated by Antoine-François de Fourcroy (1755–1809) and was read, debated and passed as the first school law in the Assemblée nationale between the 24th and 28th of September 1794 – two months after Robespierre's overthrow. It carries the title "An Elucidation of Instruction" (Développements sur l'enseignement).[15] In his preceding report (Rapport de Fourcroy[16]) Fourcroy emphasized after the end of the Revolution that one must forthwith put a stop to those elements seeking to destroy the sciences and the arts and endeavoring to wield power and control from the ruins of human knowledge. In order to help the achievements of human reason to become generally accepted it is imperative to prevent ignorance, superstition and error from reestablishing themselves. Higher education in the fields of science

[15] Langins, 1987, Annexe I., p.226–269.

[16] Langins, 1987, Annexe H, p.199ff.

and technology would thus function as a bulwark against a return to an era of despotism and arbitrary rule.

In the Rapport de Fourcroy the following teaching goals were formulated: to provide a basic education for all engineers from the various areas at a single school as well as to resume teaching the exact sciences interrupted by the Revolution. The new school was given a double task, namely to serve the Republic and to fulfill the demand formulated in 1793 to re-establish the basic sciences. Spreading the study of the exact sciences was considered to be the most forceful means of advancing the excellence of applied industrial technology. The encyclopedic chart of the sciences with its inner epistemological context served as model for the very first systematic, scientific curriculum world-wide.

De Fourcroy, chemist, member of the Académie des Sciences since 1785, co-founder of the École Polytechnique. (Nicole Dhombres, 1989, p.54)

In its founding document the School was endowed with the task to train engineers who would be capable of being employed in the various public, whether civil or military, services; their activity was not only to be directed towards defending the country and its revolutionary achievements but also towards expanding and improving French territory. On 8 February 1795 the school developed a slightly modified curriculum with regard to the inauguration of the proper course titled "Programme de l'Enseignement Polytechnique de l'École Centrale des Travaux Publics" (1 April 1795).[17] This new organizational and pedagogical guideline carries the unmistakable signature of d'Alembert and Gaspard Monge.

[17] Langins, 1987, Annexe G., p.126ff.

Conditions of Admission and the Allocation of Graduates

The following conditions of admission were formulated in the new school law of 28 September 1794: as first priority a republican attitude, then basic knowledge in arithmetic, algebra and geometry, finally an age of entry between 16 and 20 years old. The principle of the Enlightenment, that education be available to all individuals equally regardless of their social background, sex[18], ethnic background and religious conviction, was realized in practice by the school's founding fathers and the admission examinations

[18] On the first female students at polytechnical-industrial schools, cf. the chapter on the Swiss Polytechnikum in Zurich.

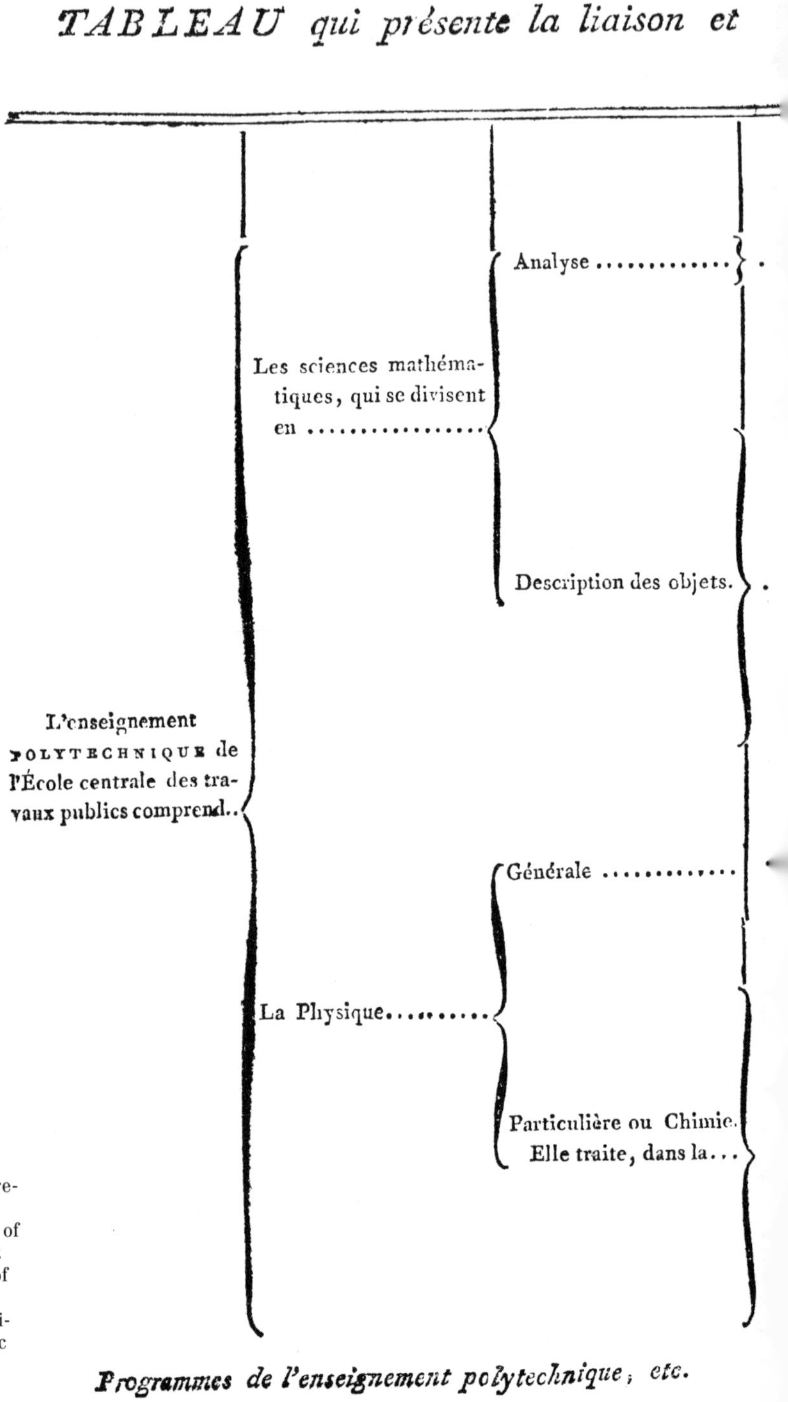

TABLEAU qui présente la liaison et

L'enseignement POLYTECHNIQUE de l'École centrale des travaux publics comprend..

Les sciences mathématiques, qui se divisent en ....

Analyse ....

Description des objets.

La Physique.........

Générale ....

Particulière ou Chimie. Elle traite, dans la...

Programmes de l'enseignement polytechnique, etc.

The scientific interrelationships of the polytechnical fields of teaching in the first teaching program of 1794 according to d'Alembert's and Diderot's encyclopedic model. (Langins, 1987, p.128–129) *l'ordre des Programmes réunis dans cette Collection.*

			Ordre des Programmes.
. . .	Appliquée à la Géométrie		Analyse appliquée à la Géométrie.
	Appliquée à la Mécanique		Analyse appliquée à la Mécanique.
. . . Susceptibles de définitions rigoureuses, d'où résultent	La Stéréotomie		Stéréotomie.
	L'Architecture.	I^e partie.	Architecture, première partie.
		IIe partie.	Architecture, deuxième partie.
	La Fortification		Fortification.
Non susceptibles, ce qui donne lieu à l'art du	Dessin		Dessin.
. . .			Physique générale.
. . . Première partie	Des substances salines		Chimie, première partie. Des substances salines.
Deuxième partie	Des substances végétales		Chimie, deuxième partie. Des substances végétales.
	Des substances animales		Chimie, deuxième partie. Des substances animales.
Troisième partie	Des Minéraux		Chimie, troisième partie. Des Minéraux.

were distributed across all of France and in 22 provincial cities (Geneva and Brussels were included later). This prevented the development of a concentration on wealthy circles in and around Paris and facilitated a mixture of territorial, social and cultural backgrounds.[19] Subsequently, the experts were to propose suitable candidates and from the entire national list 400 students were finally to be definitely accepted to the École centrale des travaux publics.

[19] The statistics prove this; cf. Shinn, 1980.

The law planned to allocate successful graduates of the new school according to their abilities to continuing special schools belonging to the military or civil government sectors (for instance the École du Génie militaire in Metz, École des Ponts et Chaussées in Paris), however on the condition that they would be called in when necessary for special service to the Republic, which played an increasingly important role as a consequence of the napoleonic expansion of power.

The Teaching Staff

The school's first faculty comprised nine professors, seven assistants and three masters in all. The teachers belonging to this first generation were: La Grange and Laplace as mathematicians, Prony in mechanics, Monge and Hachette in geometry and stereometry, Delorme and Baltard in engineering design and in architecture (Travaux civils), and finally the chemists Fourcroy, Vauquelin, Berthollet, Chaptal and Guyton de Morveau. Because the new school was able to engage the most renowned scholars as teachers it quickly became one of the leading institutes in the world. An entire staff of highly qualified draftsmen and sculptors were responsible for providing the materials in those subjects where drafting tables and models were in use; one of these was also Jean-Nicolas-Louis Durand in the field of architecture. These experts also provided support in class.

The School Organization

The school organization was made up of an administration and a School Council. A director was general administrative supervisor, took care of the accounts and was responsible for the school materials and equipment. He was also chief of personnel. A vice-director organized the school's security together with three assistant managers, who supervised the studios in all three student levels. It must be noted here that in alignment with the student's political trends, disturbances continuously broke out resulting in disciplinary measures – expulsion included. Often political activities from outside were also brought into the school.

The School Council was responsible for the classes and for making sure that the teaching activities functioned well, for promoting the excellence of science and the applied technical processes, and to inspect the teaching and class schedules. Moreover, it chose the best student works, panels and models as examples for and proof of the progress being made at school and thus also generally controlled not only the students' improvements and successes but also those of the teachers and of the school in general to satisfy the ministry of the interior and the public.

The First and Second School Building

For the first ten years the school was housed in the Palais-Bourbon building complex on the Rue de l'Université near the Esplanade des Invalides, which had already functioned as the preliminary Commission des travaux publics' quarters. Further buildings in the area were also used as institute annexes.

In a letter to the governor of the École Polytechnique of 23 March 1805, Napoleon stated his intention to move the school to another site. Too many young people had been brought together in one place and it would be better to spread them over a larger area – if, however, it was imperative for the "École" to remain in Paris and if it were not feasible to decentralize the school (a site chosen outside of Paris would have made the recruitment of professors from the circle of renowned scholars impossible), then he would prefer the Quartier Latin (!) instead of the area around the Marais des Invalides.[20] The letter ended with the request that the governor seriously address the question of moving the school. Eight months later the École Polytechnique was moved to the Collège de Navarre near the Pantheon – the crucial factor in de-

[20] Napoleon, personal correspondence of 23 March 1805.

The Palais-Bourbon, which first housed the École Polytechnique between 1794 and 1805. (Nicole Dhombres, 1989, p.90)

ciding between other possibilities was the fact that it also maintained a barracks.[21]

[21] Fourcy, 1828, p.248f.

Stages in the School's History

In the following the essential phases in the history of the École Polytechnique from the era of its inception to the year 1830 will be presented. In accord with its close affiliation to the state, the phases of the school's development will be presented along the lines of corresponding historical and political periods.[22]

[22] For a chronicle of the École Polytechnique: see Fourcy, 1828 (1987), Pinet, 1887, Callot, 1982, Belhoste, Picon et al., 1994.

The establishment phase: 1795–1799. This phase in the new school's process of development, which began with the inauguration of the regular courses with 400 students on 1 April 1795, is characterized by its struggle to achieve a privileged status and role in the recruitment of all future engineers and architects for government service, whether for the military or civilian sector. In order to not only distinguish itself from already existing engineering and officer schools but also to clearly define its role in comparison, in a constitutional supplement of 1 September 1795 this novel institution was given the name of École Polytechnique, a term which appeared in its teaching program of 8 February 1795 for the very first time.

Subsequently, on 22 October 1795, a major change took place initiated by a decisive organizational law in which the École was given a key position

The École Polytechnique at its second site proscribed by Napoleon at the Collège de Navarre, which also maintained barracks. (Callot, 1982, p.2).

in the recruitment of all future engineers and architects as well as officers; this secured their enrollment in continuing vocational and specialized schools (such as the École des Ponts et Chausées in Paris, École du Génie militaire in Metz, École de Mines and so on). As a consequence, the École Polytechnique acquired a privileged status in the network of advanced scientific and technical learning institutions in France. This privilege meant that only one path could be chosen to attend specialized cadre schools through absolving a polytechnical course of basic education. As a result, its "Privilège" was continuously being contested, yet also successfully defended, and although in the process of this century it has become much more differentiated, its status has been maintained up to the present date.[23]

[23] Grelon, 1994, p.15–57.

Plans for restructuring the lessons and disciplining school activity as of 1797 were geared toward binding the school more closely to General Bonaparte's growing military apparatus. It was the time of his campaigns into Upper and Central Italy (1796/97), of the ones planned to Switzerland and the Piemont (1798) and finally to Egypt (1798/99). The École Polytechnique and its supporters in the legislative assembly were able to prevent a restructuring thanks to the great reputation of those scholars who taught at the school and the influence they exerted in

Recording the ruins of the Palace of Karnak in Thebes, Egypt. (Napoleon, 1802, p.294f)

scientific circles throughout the world. However, the school still became subject to militarization and in the following years it was incorporated into the regime's and General Bonaparte's military projects. Professors such as Monge and Berthollet took on scientific and organizational responsibilities during military campaigns and in the conquered regions. Many students were also conscripted to serve the military. Napoleon, in his campaign to Egypt, committed besides 35,000 soldiers also a group of 167 scholars, including a corps of 39 Polytechniciens; Professors of the "École" also led the "Commission des sciences et des arts". They were to undertake archeological diggings, survey and document histori-

cal sites[24] and establish scientific institutes based on the French model. [color plate 3]

The school constitution of 30 October 1799 was to secure its existence; however it reduced the "École's" recruiting privilege by allowing those individual army units interested to determine for themselves the quantity of cadres required. Nevertheless this new organization was established with the consensus of professors, school administration and government institutions. It brought the school quieter times, provided it with a secure base and adequate means. It testifies not least to the amicable relationship maintained between Napoleon and the "Corps de professeurs" at the École. However, as a result the school became more

[24] 3,000 panels in ten folios and two anthologies with the title "Description d'Égypte": Napoleon, 1802.

closely incorporated into Napoleon's militarily organized state apparatus.

The consolidation phase: 1799–1815. The time of Napoleon – between his coup d'état in 1799 and his final deposition in 1815 – represents a time of consolidation for the École Polytechnique. The school is able to maintain its status with regard to all continuing specialized schools, the curriculum remains committed to the original model at least until 1804 (Napoleon's crowning as emperor), the school enjoys an excellent international reputation and is used as model for similar institutions in Europe and the USA. Particularly after the Treaty of Lunéville (1801), many foreign scholars visit the "École" in Paris; included among these were, for instance, Alexander v. Humboldt, Johann Gottfried Tulla, co-founder of the Polytechnical School in Karslruhe (1825), Albert de Mousson, son of the chancellor of the Republic of Helvetia and professor of experimental physics at the Zurich Polytechnikum (1855), and Guillaume-Henri Dufour, General of the Swiss Sonderbund War (1847) and co-founder of the Zurich Polytechnikum. Visiting Paris was also considered to be the starting-point for the traditional tour of Italy undertaken by prospective architects.

On the one hand, with Napoleon a time of consolidation and relative quiet followed; yet on the other hand, the law of reorganization of 16 December 1799 left its distinctive napoleonic mark upon the École Polytechnique. The law caused a discernible shift in focus towards continuing specialized schools, meaning that polytechnical basic education was to more directly serve subsequent officer training. Yet the École Polytechnique was also to supervise those schools, so that it became the dominant institute in France's scientific-technical system of education – with priority placed upon the military sector and fields of engineering. Education therefore became more efficient, coherent, and at the same time more goal-oriented and regulated.

Towards the middle of 1804 Napoleon awarded the "École" military status, designated a governor responsible for the school, ordered the students to be placed in barracks and organized into military corps. Individual professors took on ministerial posts. The

Napoleon's official visit to the "École" during the "Hundred Days", where he was enthusiastically greeted by the students. (Callot, 1982, p.53)

militarization of the school, its courses of instruction and activities and finally its move to the Collège de Navarre – which also maintained barracks – were determined in a law passed on 16 July 1804. For later critics of the polytechnical school who disparaged its tendency towards elitism, this law established the school's departure from the original model based on the concepts developed by the Enlightenment.

After the devastating losses suffered by the napoleonic army in its Russian campaign of 1812, and after the allied (Russia, Prussia, Austria and Bavaria) coalition's successful counter-campaign, France was defeated and with the allied army's entry into Paris (31 March 1814) the napoleonic era was brought to an end. Napoleon was deposed and deported to the island of Elba (11 May). The "First Treaty of Paris" followed on the 30th of May. France returned to a constitutional monarchy based on the "Charte consitutionelle". In March 1815 Napoleon returned from Elba and re-established his power with the help of sections of the army remaining loyal to him. The "Hundred Days interim" overwhelmed the École Polytechnique. The school celebrated Napoleon's return, its "champion", on the occasion of his visit to the school. Because of this, however, after Napoleon's second overthrow and his permanent banishment to the island of St. Helena in the South Atlantic (2 June) the newly restored king and his administration continued to regard the school with distrust.

The stagnation phase: 1815–1830. The Restoration phase, which lasted until the July Revolution of 1830, is characterized by a gradual dismantling of the original polytechnical model. The Restoration monarchy built upon the napoleonic foundations and instrumentalized the school and its courses of instruction for its own purposes. While Napoleon modified the school in order to recruit the cadres of officers needed for his campaigns of conquest, the new regime under Louis XVIII endeavored to re-introduce the principles of the Ancien Régime such as the economy of privilege, and to fashion the "École" into an elite school.

Portrait of the Duc d'Angoulême, new patron of the École Polytechnique. (Callot, 1982, p. 58)

With the law of reorganization of 4 September 1816 the government placed the polytechnical school under the control of a monarchist supervisory council, in which aristocrats near to the Bourbon dynasty served. The patronage awarded to the Duc d'Angoulême and its change of name into "École Royale Polytechnique" was a clear, outward signal marking the new role the former school of the Enlightenment was to take on.

It corresponded with this change in system that the original goals, namely to educate engineers and architects in the service of progress and the general welfare, to prepare the country for the modern economy and to improve the population's condition in life, were no longer a priority. High taxes and other obstacles that prevented the active participation of broad sections of the population, of trade and entrepreneurial citizens and also of middle-sized industrialists and financiers in political life, an assembly of notables ("Pairs de France"), which supported the king's hierarchized administration and the mixing of powers, all played a role in delaying the development of modern industry in France. Under the 29 million inhabitants of France only 100,000 were allowed to vote. Yet the pressures exerted by industrialization demanded that resources be made available and that free markets and liberal political structures be established.

In the course of the Restoration the polytechnical school lost its scientific-technical superiority, as the government that supported it again considered representative commissions to be of greater importance. After the reorganization of 1816 the curricular focus shifted increasingly towards instruction in theory in general, over-all subjects and a reduction in courses of application. Education and exercises in Géométrie descripive were restricted to the first-year course and it was dissolved into a principle of method that limited theory and practice. In the architecture course project exercises were reduced and replaced by courses in sketching techniques. The restructuring of these two "Pièces de résistance" of the polytechnical model of education as well as the primacy of theory – "Sciences pures" over "Sciences appliquées" – explicitly exhibits a bias against the genuine Modèle polytechnique. This is augmented by the limit of 130 set to new enrollments and the establishment of a maximum age of leave ("age guillotine"). All this only contributed to blocking industrial progress, spoiling the future prospects of many candidates and robbing them of their hope of being able to participate in modern life as architects and engineers.[25]

[25] De Comberousse, 1869.

The July Revolution of 1830 and the role of the Polytechniciens. On the eve of the July Revolution of 1830 almost all students of the École Polytechnique supported liberal opposition movements; the number of royalists remained a minority. Even the broadest sections of the population were in opposition to the government of Charles X (as of 1824 successor to Louis XVIII), who rejected the still valid Charte consititutionelle (with a minimum of basic rights), who strengthened the church and the aristocracy, and who endeavored to re-install absolutism. Charles X ignored the opposition parties' election victory (1828), intensified the royalist course (Prince de Polignac as Head of Cabinet)and finally dissolved the inconvenient Chamber (beginning of 1830). When in new elections the opposition was confirmed as the strongest party the king prevented the elected Chamber from meeting and enacted the "July Ordinances" which further restricted suffrage and liberty of the press.

As a result, the July insurrection occurred in Paris; it lasted three days and in the end brought the "citizen king" Louis-Phillippe, Duc d'Orléans – who reigned until 1848 – to power. After 1830 the modern industrial bourgeoisie, and thus also industry and trade, were able to develop more strongly and quickly – the industrial revolution in France had begun.

During the July Revolution the Polytechniciens sided as expected with the insurrection; moreover, they took charge at strategic sites, and as fearless and decisive young officer candidates they contributed to bringing about the Monarchy's defeat.

Many eye-witness reports confirm the outstanding role that the group of about 60 students from the "École" played. After the successful overthrow the courageous polytechnical students were overwhelmed with congratulations from all over France and poems were published in newspapers in their honor.[26] (color plate 4)

[26] Pinet, 1887, p.162–165.

The Polytechniciens in the July Revolution. On 27 July 1830 the unrest began. Soon after the newspapers were suppressed the "Boulevards" rose in uproar, the first barricades were set up and the insurrection mobilized the students. Towards the evening the "École" was also mobilized. However, on the following day a royal decree closed the school and the school's direction sent the students home. About 60 Polytechniciens decided to take part in the revolt and took position at the front lines of the insurgent population. On 28 July, the second day, they fought at different sites and led the rebel groups. On 29 July the decisive turning-point occurred after most of the quarters had been taken, the "Garde royal" dissolved and their command given to the elderly Lafayette. Strategic sites such as the Luxembourg horse stalls, the artillery depot and arsenal were brought under control. And everywhere the Polytechniciens were met with wild applause by the population –they were the celebrated leaders and heroes of the successful rebellion. Towards midday on the 29th of July the revolution was decided –the king's bastions fell. Already in the evening of the same day a new government was established, which represented the union between the populace and the army and installed the former chief of the Garde royale, Louis-Phillippe, Duc d'Orléans as citizen king.[27]

[27] Pinet, 1887, p.139–162.

The July battles in the streets of Paris and the Polytechniciens' major participation in them were also registered internationally, which is evident in the laudatory reactions coming from Belgium, Italy and Switzerland. A letter of solidarity was written by students of its associate school of West-Point, New York (1 October 1830), in which they were congratulated for their heroic and decisive battle in re-establishing the constitution, constitutional government and civil rights.[28]

[28] Pinet, 1887, p.162–165.

The Making of the Polytechnical Educational Model

Committed to the ideas of the Enlightenment, the founders of the École Polytechnique were convinced that a higher, comprehensive and scientifically based system of education for engineers was capable of greatly influencing social progress. The sciences were therefore not only to be taught and maintained in academic isolation but also employed as the basis for the practice of technical and industrial benefits. With this aim they developed a specific educational model connecting theory and practice.

It was the first time that an institutionalized school for architects and engineers directly combined instruction in theory with practical application according to concrete and socially useful tasks, and that these were adapted according to the demands of the time. This novel crossing of theory and practice that was to define the polytechnical tradition of education is a product of the Lumières' school of thought. The didactic innovations established by Monge and Durand formed the core of the Modèle polytechnique's special and revolutionary character.

On the problem of theory and practice of the Lumières. The philosophy of the Lumières did not exhaust itself in theory but sought to become manifest in social practice. Revolutionary achievements such as human rights and constitutional government are products of this. One of the first to have a direct influence on the times as a philosopher and historian was Voltaire. His opposition to the arbitrary practice of justice in the Ancien Régime showed courage and far-sightedness; this is exhibited, for instance, in his critical account of the "Calas case". Condorcet considered him to be exemplary in his social commitment. D'Alembert and Diderot's "Encyclopédie" was also compiled with the intention to make available to the interested public information on the state of the sciences and of experience in order for this to be directly applied in practice in workshops, studios and schools, and thus to raise the level of craftsmanship, trade and industry. When

Turgot, who a member of the circle of Enlightenment scholars, was appointed minister of finance under Louis XVI in 1774 the "philosophes" interpreted the king's decision as a liberal shift and the political realization of their principle postulates. Turgot subsequently appointed Condorcet as his advisor and installed for example a commission to develop French canal construction in order to expand inland shipping; besides Condorcet, d'Alembert and Bossut, other members of the Académie des Sciences were also engaged as experts. Condorcet, in his function as "Inspecteur des Monnaies", was further responsible for standardizing the French system of measures, weights and coins, which corresponded to the demand of the Enlightenment for a free and no longer privileged access to the market. This and other activities illustrate the project of "applied philosophy" as realized from 1774–1776 during Turgot's short term of office.[29]

[29] Voltaire, 1975 (1993); d'Alembert, 1751 (1975); on Turgot, cf. Condorcet, 1786; on Condorcet, cf. Badinter, 1989, as well as Arago, 1854.

Portrait of Gaspard Monge. (Pothier, 1887, p.371)

Gaspard Monge and the Géométrie descriptive

Gaspard Monge belongs to the formative individuals of the polytechnical method of teaching and learning. He originally formulated the Géométrie descriptive, which was to become the core science in the canon of subjects maintained by the polytechnical school. In the course of his work, first as assistant and collaborator, then as professor of physics and mathematics at the École du Génie militaire in Mézières, he developed the theory of "descriptive" geometry as a systematic method for solving and depicting practical problems of geometry three-dimensionally. Géométrie descriptive is, however, more than just a technical method. In his textbooks, which were partly published by his later collaborator J.N.P. Hachette, not only are practical problems shown and many examples of how to apply problem-solving methods presented, but also detailed descriptions of the theoretical principles are included to accompany each step in the problem-solving process.[30]

[30] Monge, 1801; Hachette, 1822.

Monge's Géométrie descriptive is a dynamic and flexible method that was applied to all then relevant disciplines of engineering in polytechnical education and in praxis: as a way to pose a problem, to geometrically record static or dynamic processes, to describe the solution process and project solutions in bridge, road and canal construction, for harbor facilities and fortifications, for architecture and urban planning, machine construction and hydraulic works, manufacturing facilities that process raw materials and for processing metals and chemicals, etc. In an introduction to the "Journal de l'École Polytechnique" Monge emphasized that modern engineers use the science of Géométrie descriptive not only as an instrument of communication but also in order to solve the most different assignments demanded by the time, and that as a consequence it has developed into a kind of modern language with which also less comprehensible problems could be mastered.[31]

[31] Monge, 1794, p.3.

The "stone section" was the most elementary example used by Monge. In addition to the silhouette projection and the circular stairway to be treated further on, the construction and projection of double curvatures and planes, a problem to which Monge accorded great importance and to which numerous mathematicians of the time applied themselves, will be touched upon and explained. The problem arose, among other things, in the context of recording the kind of complex sequences of movement that took place in industrial plants and methods of processing. In this respect movement is recorded in instances and projected onto vertical and horizontal planes; an initially unfamiliar appearance, such as a movement or even the shape of a contour are formed by or reduced to mathematically defined lines, planes and volumes, and with the aid of projectional techniques are presented in detail.[32] This will be illustrated by an example from Auguste Perdonnet's railway textbook.[33]

[32] Monge, 1801, p.106ff. as well as Planche XXIII.

[33] Teacher at the École Centrale des Arts et Manufactures, founded in 1829, where Théodore Olivier, a student of Monge, taught Géométrie descriptive.

The Géometrie descriptive, which was first taught by Monge and Hachette and later, after Monge's dismissal in 1816, by Dominique François Arago, played a prominent role in the educational program and curriculum of the polytechnical school. It was not only the basic subject but also a specialized science that permeated the practice courses and project work of the higher class levels; it also came to characterize the entire polytechnical system of education, which was also soon to be introduced in other countries and schools, as for instance in the industrial École Cen-

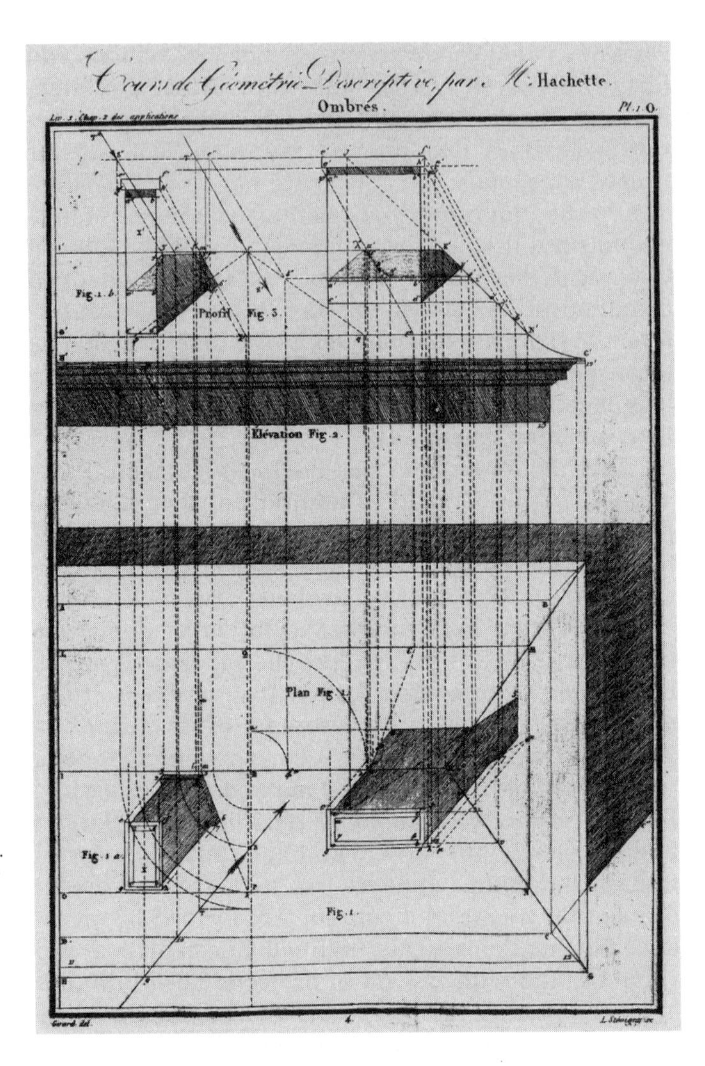

The structures illustrated in Monge's textbook each treat a particular site, perspective and moment and therefore serve a methodical approach of space and time and facilitate a transference to other, unknown problem areas. Here a silhouette projection and a circular stairway are used as illustrations in the textbook compiled by Monge's colleague Hachette. (Hachette, 1822)

trale des Arts et Manufactures (in the following as "École Centrale").

The Educational Concept

The polytechnical educational concept rested on two pillars: an organized system of subjects and the principle of practice courses. The two major disciplines of the sciences of the Enlightenment, mathematics and physics, were also the focus of study in polytechnical education for all fields of engineering and for architecture. The scientific interrelationships between both

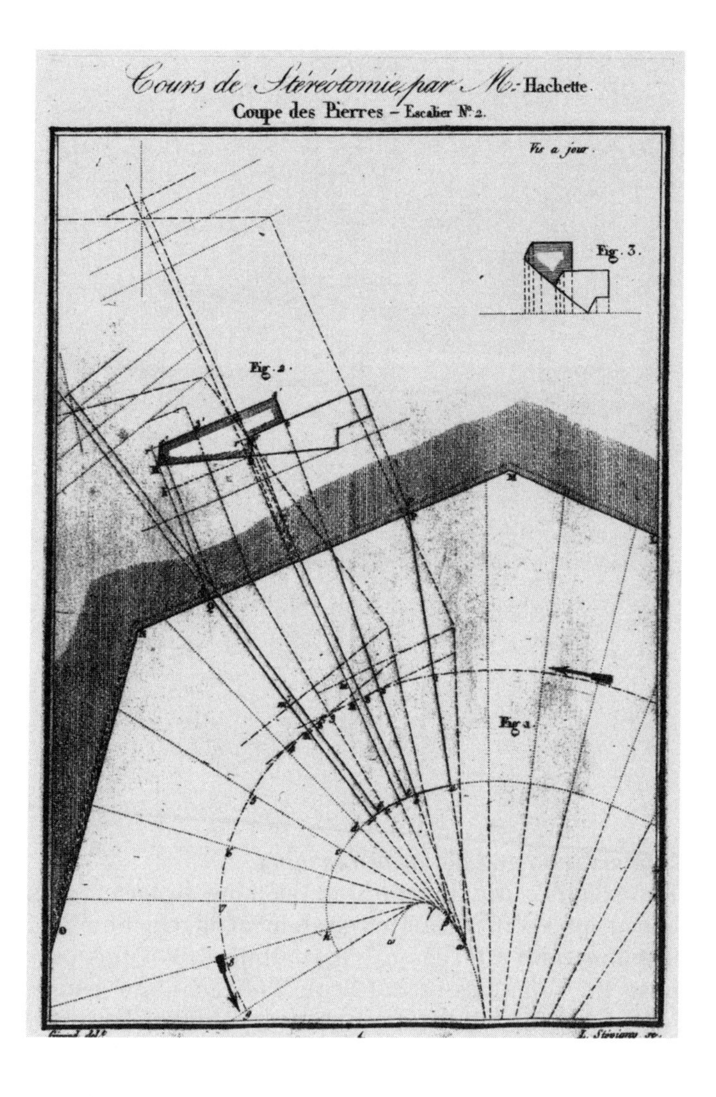

basic disciplines and the transfer that was to take place in the practice courses (Cours d'application) of engineering and architecture were methodically transmitted by the Géometrie descriptive.

The course of study was initially divided into three year-courses: stereotomy in the first, architecture in the second and fortification in the third. Drawing, painting in water-colors and modeling were added as a further discipline for students to learn how to record and reproduce basic aesthetic elements of nature and human activity.

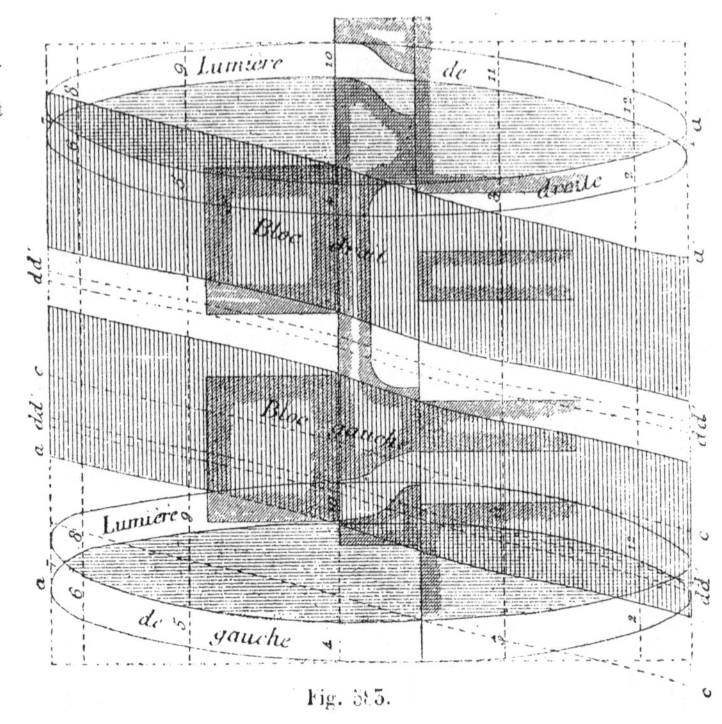

A section of a piston movement of a locomotive and a demonstration of the machine parts' different lines of movement. (Perdonnet, 1860, p. 500)

Organization and Forms of Teaching

The school's organization or teaching program was based on a differentiated system of instruction and learning with a focus on "Perfectibilité". For this purpose the founders of the Ècole Polytechnique established a didactic framework consisting of lectures,

The Dutch scientist Brugge visits a laboratory of the École Polytechnique in 1798 and is impressed by the high standard of its facilities (N. Dhombres, 1989, p.95)

studio courses with exercises, projects and testing. A combination of basic instruction in theory with a diverse system of practice and concrete experience forms the didactic core of the polytechnical educational model.

While lectures took place in the large auditorium (Amphithéatre) and served to impart theoretical knowledge, students spent the rest of the time in drawing studios, laboratories and workshops. Magistral lectures also were an opportunity for the students to become acquainted with the teacher, who, in his role as an exemplary personality led his students through the subject matter and provided an example of scientific thought and eloquent presentation. In addition, in using their textbooks professors such as Monge or Durand were able to open up new perspectives, explain terms, establish correlations and explain the material by using concrete examples.

Assistants prepared the lectures, compiled the textbooks (in cooperation with professional draftsmen) and during practice courses answered the students' pertinent questions arising from the lectures. Finally tutors – older students or graduates (Répétiteurs) – were responsible for organizing the classes, for matters concerning discipline and personal questions. Also employed was a class leader (chef de brigades), who in his function as a role model was responsible for managing and guiding a brigade – as the school body was divided into brigades of 20 students each – and saw to it that the "daily agenda" or class schedule was observed.

The didactical complementarity between the studio and experimentation. The founders of the polytechnical model of learning and teaching emphasized in the "Elucidation of Instruction" (1794) mentioned above that the combination of spectacular and attractive physical and chemical experiments in the laboratory with more placid and abstract mathematical and geometric operations in the studio and auditorium provided not only an important variety in class, but also that the interrelationship between these two different didactic methods served to encourage interest and in a sense created a "double" motivation.[34]

[34] Quoted from: Fourcy, 1828 (1987), p.42, [notation 1].

Individual examinations and tests were included in the spectrum of instruction forms used at the school. (Callot, 1982, p.458)

The polytechnical mode of teaching was characterized by considerable personal interaction between the student and professor, Répétiteur or tutor. Class instruction, which was introduced for the first time as a program at the École Polytechnique, was thus based on the model of a solid social group as a learning community and on instruction by professional teachers as exemplary individuals. At the same time the students' learning ability was continuously being stimulated through exams or tests (Interrogations) given at short notice.

The precise formulation of problems, of program requirements and judgment criteria facilitated a practical approach to problems and prove that "enlightened" teaching methods were applied so that all students would be able to master their tasks; at least in its initial phase the "École" was not concerned with becoming an institution for the talented elite.

Field work was aimed at not only providing useful experience and practical knowledge pertaining to different vocations; it also endeavored to offer the students a healthy contrast to the stuffiness of the study room. Trips and excursions outside the school took place to complete specific assignments or self-appointed tasks, which were subsequently to be described in the form of notes or reports. These exercises were to spur the students' ambitions, expand their horizons and cultivate their taste.

The didactic model. A differentiation of teaching and learning methods follows the idea that in combining instruction in theory during lectures given by professors (Cours) with verbal examinations (Interrogations), written exams (Concours), practical exercises (Travaux), research (Opérations), experiments (Manipulations) and concrete experience (Visites et Excursions) it would form a comprehensive system of teaching and methodology in which different mental processes such as learning, understanding, repetition, recall, application and interpretation – in modern terms of learning theory known as primary strategies of learning – are interconnected and organized.[35]

[35] Cf. in detail the basic description of the learning process from the perspective of learning-psychology: Steiner, 1997 and 1999.

Utilizing Teaching Materials

The faculty met in regular scientific meetings, which were recorded and found expression in the school's novel use of teaching materials, the first of its kind ever to be used. On Monge's impetus the School Council published a "Bulletin de l'enseignement" with a large edition of 3,000 copies. This bulletin contained the results of expert discussion, individual treatises by professors as well as teaching programs and treatises on teaching methods, making the circle of scholars renowned throughout France and the world. Personages from outside the school were also invited to School Council debates, so that it acquired the importance of an academic site par excellence. Its meetings represent a milestone in the post-revolutionary establishment of scientific activity in France and presented itself – as a chronicler of the school, Antoine Fourcy, states – as a "Compagnie savante" in the eyes of the same suspicious group who a few years before had liquidated the academy and all scientific corporate bodies, as they had stood in contradiction to their postulate of equality.[36] The Bulletin, which already a year later appeared as the "Journal de l'École Polytechnique", soon became a leading specialized journal worldwide and published articles by Monge, Fourcroy, Lagrange, Laplace, Prony, Malus, Ampère and many others.

[36] According to Fourcy the group of Jacobins; Fourcy, 1828 (1987), p.67.

The general impression is given of an extensive "culture" of teaching materials established and maintained by the École Polytechnique. It corresponded completely with the spirit of the Enlightenment in disseminating knowledge and experience to encourage learning and practical application. Because every teacher was obliged to base his teaching on scientific reasoning and to present this formally, the entire body of teaching material appears to us today as a "Tableau" – a systematic and encyclopedic presentation of the arts of engineering – or as a demonstration of the "state of the art" of polytechnical education of that time.

Teachers and Students

The original nature of polytechnical education was actually that of an "education of encouragement". According to Monge, teachers not only had the duty to organize their teaching activity, but also to encourage their students in the acquisition of knowledge, to challenge them to new heights of qualified achievement – which every student enrolled was considered capable of – and to spur their ambition and enthusiasm. This basic teaching attitude promoted the students' interest in their work, their identification with the problem to be solved and a sense of responsibility toward their own work. Moreover, by introducing and organizing an environment of competition in class one sought to achieve a perfectibility in the students' achievements. Thus, as of 1806 two of the best student works were selected by a jury at the end of each semester and presented in the school's exhibition hall. This inspirational collection increased from year to year.

In accordance with the definition of educational work formulated, polytechnical teachers assumed an important teaching responsibility towards each individual student, who in his later profession was to also make a positive contribution to cultural and social development. Furthermore, within the class community situations of social interaction equally prepared the students for similar situations in their later working life.

The class community system was particularly effective in the practice courses, drawing studios, workshops and laboratories. (Callot, 1982, p.314)

The question naturally arises what effect this model of education – then still a pioneer achievement – had on the students. Individual teachers were honored and described by former students, which provides us with an insight into the class environment of the "École" of that time. Although not yet researched in depth, from observations and descriptions available to us we can observe new phenomena emerging from the introduction of a systematic form of instruction (ever since the Enlightenment), from the conscious application of learning processes and the differentiated use of diverse forms of teaching and learning.

What can be determined in general is that teachers and students were indefatigably engaged in achieving student success in learning, in increasing their vocational qualification and improving their character. An example is Charles Dupin, former student of Monge (diploma in 1801), with his description of his former teacher – how Monge reacted patiently and paternally to difficulties and problems the students' brought to his attention after his lectures; how he corrected mistakes with forbearance and with the attitude that they could be avoided the next time; how he praised correct work, not in the sense of an absolute-seeming value judgment, but rather as encouragement for further progress. Even his manner of lecturing and of instruction left a lasting impression upon his students.

Dupin writes in his small monograph that Monge had the unique and unmistakable ability to explain complex and abstract subject matter in such a way as to be easily remembered – lively and eloquently, the structure of his statements varied by descriptive hand movements and underlined by the use of different sentence lengths, tenses and emphasis. In his manner of instruction, Monge stimulated the students' different senses and could command their complete attention – the life-elixir of education! Dupin thought of Monge's teaching as an art.[37]

[37] Dupin, 1819.

Durand, professor of architecture, is also remembered as an encouraging personage. His former German student Clemens Wenzeslaus Coudray recorded in his memoires: "I was now able to address my studies completely and without care and made such progress that one day, as Professor Durand was going through his students' compositions and while inspecting my sketch he said the following encouraging words: Coudray vous serez architecte. My beloved teacher's praise and favor, of which I had the luck of being able to continuously increase, if possible doubled my efforts to match it." In reference to the brilliant report that Durand awarded him on the occasion of his departure from Paris in 1804, Coudray states further: "This report, from one of the greatest and famous architects of the recent age, is still today my greatest pride, and throughout my whole life I will try to prove worthy of it."[38]

[38] Trans. from Coudray, Erinnerungen, cit. from Szambien, 1984 (Doc. 20, p.162f.).

"Le Modèle Polytechnique"

The founders of the École Polytechnique developed a completely new concept of education. One important educational innovation was the introduction of regular class instruction, in which the students were divided into permanent learning groups maintained throughout all subjects and year courses. A novelty with respect to teaching and methodology is also the scientific foundations upon which all engineering and architecture subjects were based as well as the practical orientation given in exercises and project work. The Géométrie descriptive imbued all courses and fields of engineering and architecture with a connection between theory and practice. It was originally intended to also establish a relationship to industry. However, after 1816 the École Polytechnique abandoned this approach, which facilitated a new industrially oriented school of architecture and engineering to become founded. As the École Centrale, it replaced the polytechnical school in its role as one of the world's leading educational institutions.

Teaching Principles of Architectural Engineering at the École Polytechnique

In contrast to traditional vocational training for architects provided by the Académie d'Architecture and the architecture courses at the École des Ponts et Chaussées of the Ancien Régime, the newly founded École centrale des travaux publics (1794) introduced for the first time a regular and systematic construction course for architects and engineers involving class instruction and structural studies divided into year-levels.

The modern curriculum for architecture established by this school was the first one of its kind ever to be introduced and reflected the needs and the demands made upon design by the post-revolutionary era. The young Republic was in need of engineers and architects capable of addressing the new tasks of the time posed by industrial development: covered markets, infrastructures, housing complexes with heating, factories, hospitals, schools; as of 1835 railway stations and towards the middle of the century story-height construction for national and international exhibitions.

The First Course in Architecture by Lamblardie and Baltard

Louis-Pierre Baltard (1764–1846) was the first professional teacher of architecture at the École centrales des travaux publics; he taught the construction course together with Jacques Élie Lamblardie during the preliminary Cours révolutionnaires and taught architecture at the newly opened school on 1 April 1795, which on 1 September of the first year was renamed the École Polytechnique. The new school's syllabus of September 1794 contains a detailed section on content and structure of instruction in architecture.[39] Baltard remained as a teacher until the class of 1796/97. At the same time he also held the office of city architect of Paris. His son, Victor Baltard, was the later builder of the "Halles centrales" in Paris (1854–1857 and 1860–1866, together with Félix Callet; torn down in 1971).

[39] Facsimile print in: Langins, 1987, Annexe G, p.146–162.

Lamblardie, who studied at the École des Ponts et Chaussées and specialized in port facilities, took over for the time-being two thirds of the course in architecture and taught road, bridge and canal construction as well as the planning and designing of port and shipyard facilities (Architecture, première partie). Baltard taught the actual architecture course (deuxième partie). One of his co-workers on this subject was none other than Jean-Nicolas-Louis Durand, who as draftsman had supported the architecture course since the beginning of the preliminary courses in December 1794.[40]

[40] Szambien, 1984, p.62.

Lamblardie's teaching activity did not last long. He belonged, as did Monge, to the closest circle of founding individuals of the École centrale des travaux publics, and was its first director; yet his office lasted one school year, as he also supported (since 1793) the ailing director of the École des Ponts et Chaussées, Jean-Rodolphe Perronet, and after his death already in September 1795 assumed his post at the same time. However, Lamblardie also died in 1797, which meant that the architecture course had to be completely reorganized.

As "Professor adjoint" to Lamblardie, Baltard taught building construction and also special building forms such as monuments and facilities for popular events, which then and since the Revolution had become an integral part of the new society. In the "Journal de l'École Polytechnique" of 1796 he published an interpretive guide to his teaching principles (Architecture. Texte de l'Enseignement de cet Art), which is to be considered the first major document on the new form of polytechnical architectural education before Durand.[41]

[41] Baltard, 1796.

Besides teaching principles of architectural engineering, which were based on characteristic examples from the history of architecture, Baltard's course also addressed problems of material strength, climate and soil conditions, with damage to building as a consequence of not following principles of construction and techniques in projection and drawing. Also treated was the correlation between building typology and choice of material, among other things by examining the differences between public and private buildings, administrative and living facilities. Not least, Baltard lectured on the advantages of rural building, in which in his opinion civilization's natural beauty is reflected. Statements such as this particularly signalize the "profane" orientation of the new teaching in architecture and reflect the modern demands that a new public emerging from the last decades of the Ancien Régime began to place on the art of building.

What was also novel is the fact that the polytechnical school offered education in architecture parallel to the one available at an academic institute. The Académie d'Architecture continued in post-revolutionary times in the form of the new École des Beaux Arts, which was maintained separately from the Institut de France and which, like the École Polytechnique, was founded on 25 October 1795.[42]

[42] Pevsner, 1973, p.199f.

Durand's Polytechnical Principles of Architectural Education

After Lamblardie's death in 1797 as well as after Baltard's departure the school engaged Durand as "Instituteur d'architecture" to lead the polytechnical course in architecture. It is remarkable how Durand, who came from an academic background himself, was able to be of formative importance in providing the polytechnical model of architectural education with its revolutionary character and in preparing the wayy for a later industrially expanded education in architecture.[43]

[43] The observation follows Szambien, 1984.

On Durand's Biography

Jean-Nicolas-Louis Durand (1760–1834) grew up in modest conditions in the Parisian quarter of Sainte-Geneviève and went to school in the Collège Montaigue nearby, which stood at the later site of the Bibliothèque Ste. Geneviève built by Labrouste. According to his biographer J.-B. Rondelet, Durand was very impressed by the Panthéon in the area, which was to inspire his architectonic and didactic work. From early on he worked as an architectural draftsman. When he was 15 he entered the architectural office of Pierre Panseron, student of Jacques-François Blondel from the Académie d'Architecture, and at the same time opened a small drawing school. A year later (1776), he changed over to the studio of Étienne-Louis Boullée, also a graduate of the Académie d'Architecture and student of Blondel, where for the first time he was paid for his work. There he had the opportunity to cooperate on and learn from major building projects such as the Hôtel Royal des Invalides (1778), the École Militaire (1780–1782), the Opéra (1781) and the Métropole (1781/82).

Portrait of Durand by Dissard. (Szambien, 1984, p.213)

Parallel to this Durand also studied at the Académie d'Architecture until 1782 and attended courses by Perronet and Leroy. In the years 1779 and 1780 he participated in a competition for the Great Prize and both times was awarded the second rank of Grand Prix de Rome for his projects; in the person of Boullée he found his mentor and received expert support.

Durand received his intellectual and creative inspiration from the tradition maintained in the Académie Royale d'Architecture, founded in 1671 by Colbert in the reign of Louis XIV, from the course in architecture taught by J.-F. Blondel, which was dedicated to the thought of the Enlightenment, as well as from subsequent teachers such as Boullée who continued the tradition of rational architectural education.[44] In addition, his teacher Perronet, in his function as professor at the academy as well as at the École des Ponts et Chaussées and a leading engineer and builder of, for instance, the Pont de Neuilly and the Pont de la Concorde over the Seine in Paris, exerted a major influence on Durand and on the structurally oriented methodology that he developed later.

[44] Picon, 1988, esp. Chaps. 3 and 4. On Boullée from a historical and architectural-theoretical context cf. in detail Vogt, 1969 and 1987.

In Boullée's studio Durand became acquainted with Pierre-Thomas Thibault, with whom he founded a studio in 1793; in the two-year reign of the Jacobins (1793/94) the studio participated in many competitions and enjoyed spectacular success. The parties and celebrations of that revolutionary phase represented an extraordinarily unique chance for architects to experiment, even though the assignments were merely propagandistic in nature; as a consequence of Robespierre's overthrow, however, the projects never were realized.[45] Durand and Thibault's studio was able to completely catch the spirit of the times and reflected the predominating desire for radical, emotive and symbolic forms of expression as well as for references to antiquity in design.

[45] For the over 120 competitions beginning in April and May of 1794 Durand and Thibault submitted 15 projects, of which 12 were given first prizes!

Durand's biography as well as the success he achieved and the renown that he received through revolutionary competitions led to his recommendation for teaching at the École Polytechnique.

Durand's Architecture class

The school administration commissioned Durand in 1797 to organize the course in architecture within a standardized curriculum and also to completely reformulate the basic teaching principles and the basics of architecture theory. From the very beginning Durand maintained that in the short period of time given, students could be efficiently and effectively educated and prepared as architects, however not as "complete" architects.

In 1799 he formulated for the first time his principles of architectural education in the "Programme du cours d'architecture". In the first years instruction comprised of 30 lessons and 13 exercises.[46] Between 1806 and 1811 the curriculum in architecture for the first-year level included 58 one-hour lectures, 25 exercise sequences lasting three and a half hours each and 8 examinations lasting ten hours each. In total, 225 lessons were dedicated to architecture in the first year of study. For these courses Durand compiled his famous teaching material "Précis des Leçons".[47] After the major reform of 1811 the number of lectures was reduced to 38 hours while drafting work was raised from 87 to 150 hours.

[46] Szambien, 1984, Doc. 13, p.156. Basic work on Durand's architectural theory include among others Germann, 1980 and Villari, 1990.

[47] Durand, 1802–1805.

Class Organization and Teaching Activity. Durand's instruction in architecture was divided into three categories: lectures (Cours magistral), exercises (Travail graphique) and examinations (Concours). This system of organization corresponded with the polytechnical tradition initiated by Monge. Lectures introduced the subject matter to be pursued in the practical exercises and in the drawing studios. The examinations that took place at the end of each year and at the end of the complete course of study were given in the form of projects, and were to show whether the student had understood the learning material and whether he was able to apply this to novel assignments.

Durand's lectures followed the Précis teaching material, thus establishing its importance as a guiding principle. The buildings Durand used as examples in his textbooks were brought up to date from one year to another in order to include the most recent Parisian constructions in his classes, such as the Pont des Invalides by Navier, the Paris Exchange or his own projects such as the Place de la Concorde, etc.

Student notes such as the valuable 70 pages from A. Fransoz' notebook from the year 1823 (now in the archives at the École Polytechnique) show that Durand often made drawings and typological sketches on the board, and added statements and propositions that underlined and illustrated the subject treated in the lectures.[48]

[48] Fransoz, 1823.

A page from Fransoz also contains a description of the seven examinations that lasted one week each and took place during the summer months. These en-

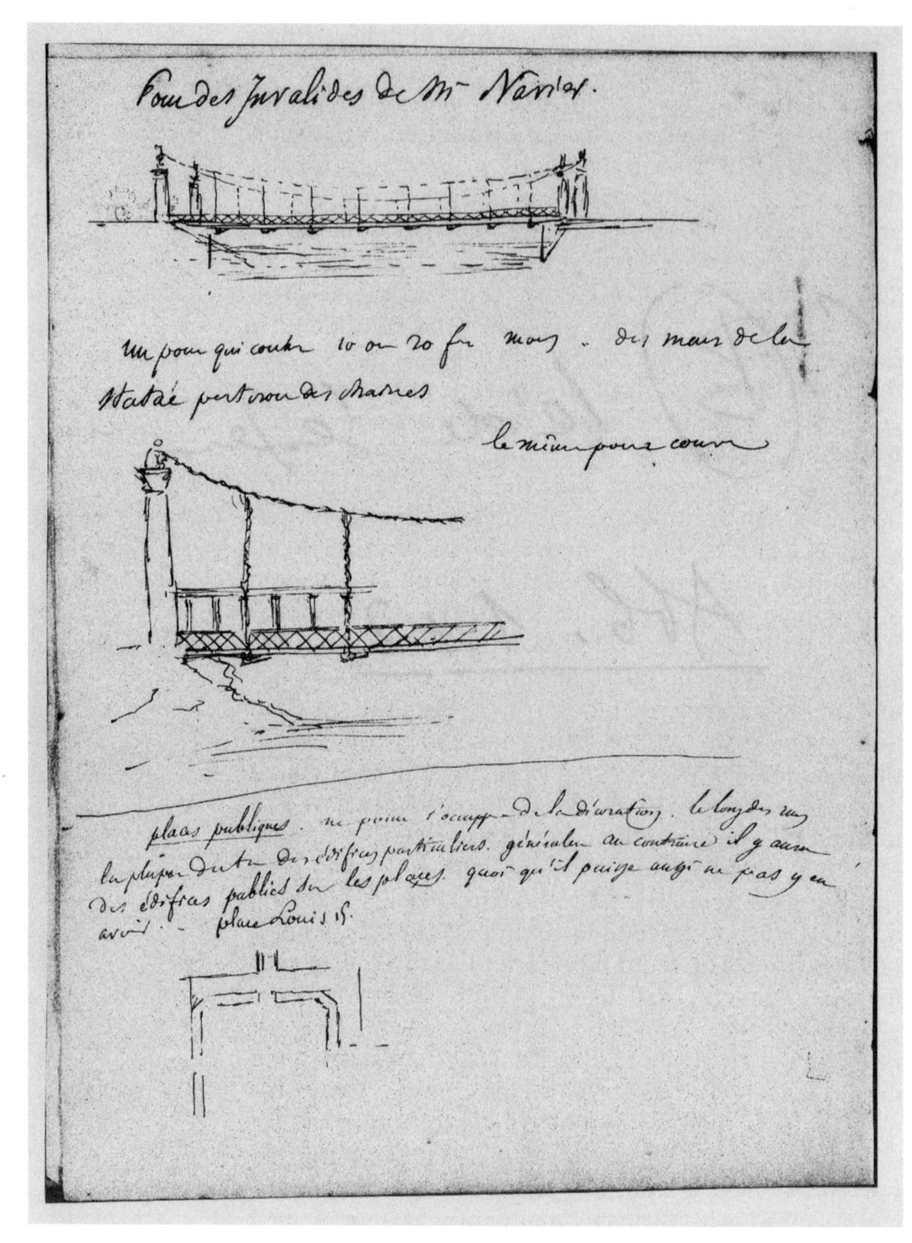

Notes taken by Fransoz, a student in Durand's architecture lectures of 1823, show the complexity of themes treated: the Pont des Invalides by Naviers and a early chain suspension bridge as well as a rare perspective, the street front of the Via Po in Turin. (Fransoz, 1823)

Ensemble des villes, villages &c. ~~on n'en pas~~ le plus souvent on n'est pas maître de la disposition générale à cause des localités.
— nous nous occuperons seulement des moyens de communications.
rues, ponts, places publiques. — objets qui sont indispensables
la communication doit être la plus courte la plus facile et la plus commode. — 1re condition, ligne droite. — pour les autres conditions. pour éviter les boues, trottoirs. pluie et soleil. — placer au dessus des trottoirs des véritables ~~routes~~ portiques. — Il faut que la ville soit saine. en supposant tout cela fait il faut prévenir les accidents terribles des incendies. Dans toutes les rues de la plupart des villes toutes les maisons sont contiguës. ce qui empêche la circulation de l'air. il serait donc avantageux de les séparer par des intervalles. mais avec cette disposition on prend bien plus de terrain? maison évitera à la fois l'insalubrité et les dangers des incendies. ~~mais reste à savoir que dans un terrain donné~~ et on pourrait alors augmenter le nombre des rues sans inconvénients. — La différence des fortunes et des conditions entraîne la dissemblance des habitations.

Symétrie et variété sources différentes de toutes les beautés de la nature et de l'art.

les portiques réunissent les maisons.

Les portiques devraient être par le même usage dans toute la ville, donc d'une construction uniforme. raisonnement contraire pour les habitations. décoration accessoire. statues si entre ces espaces des jardins l'effet serait magnifique

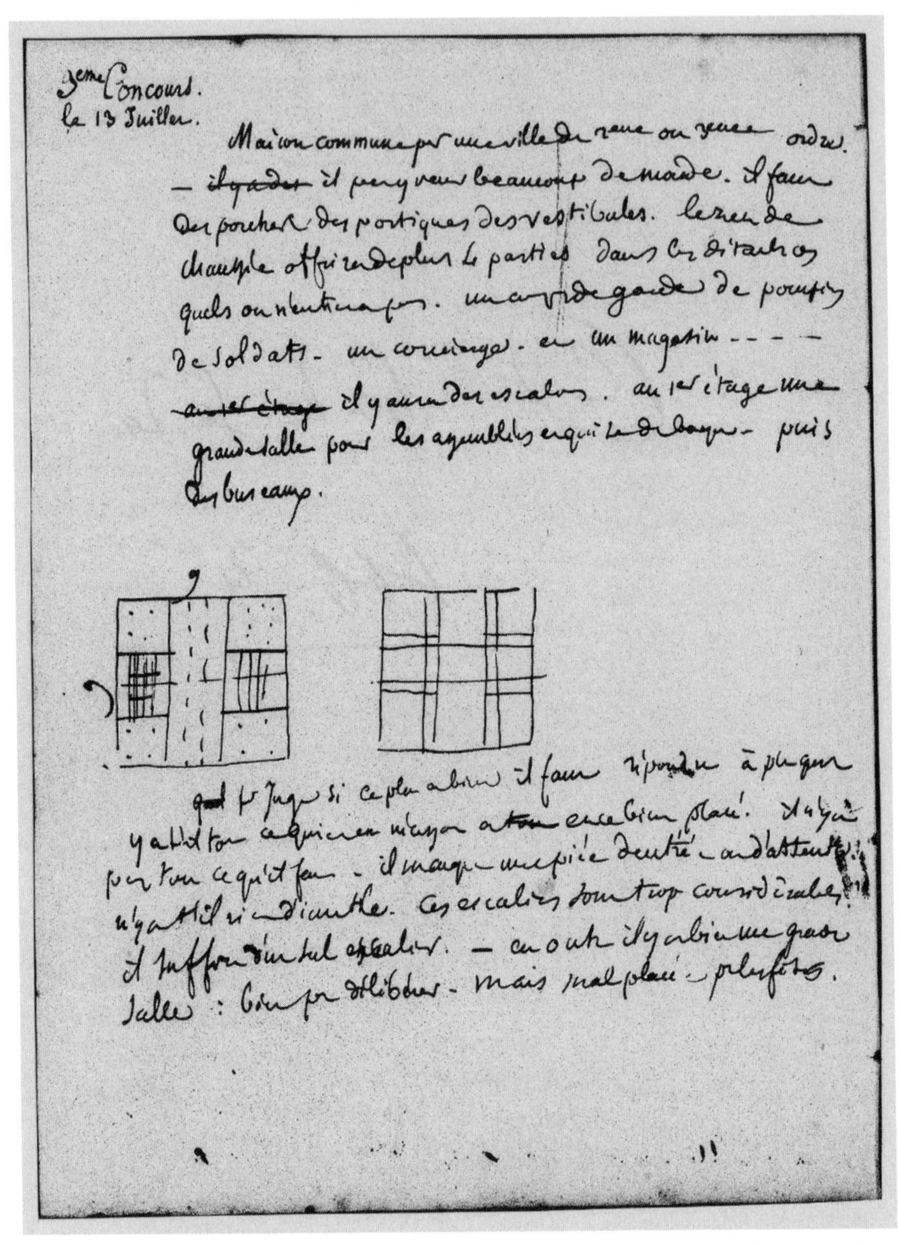
3eme Concours.
le 13 Juillet.

Fransoz' notes on the 3rd examination that took place on 13 July 1823; it concerns the design of a multi-storeyed building with different functions in a middle-sized city, which was to be completed within a week. (Fransoz, 1823)

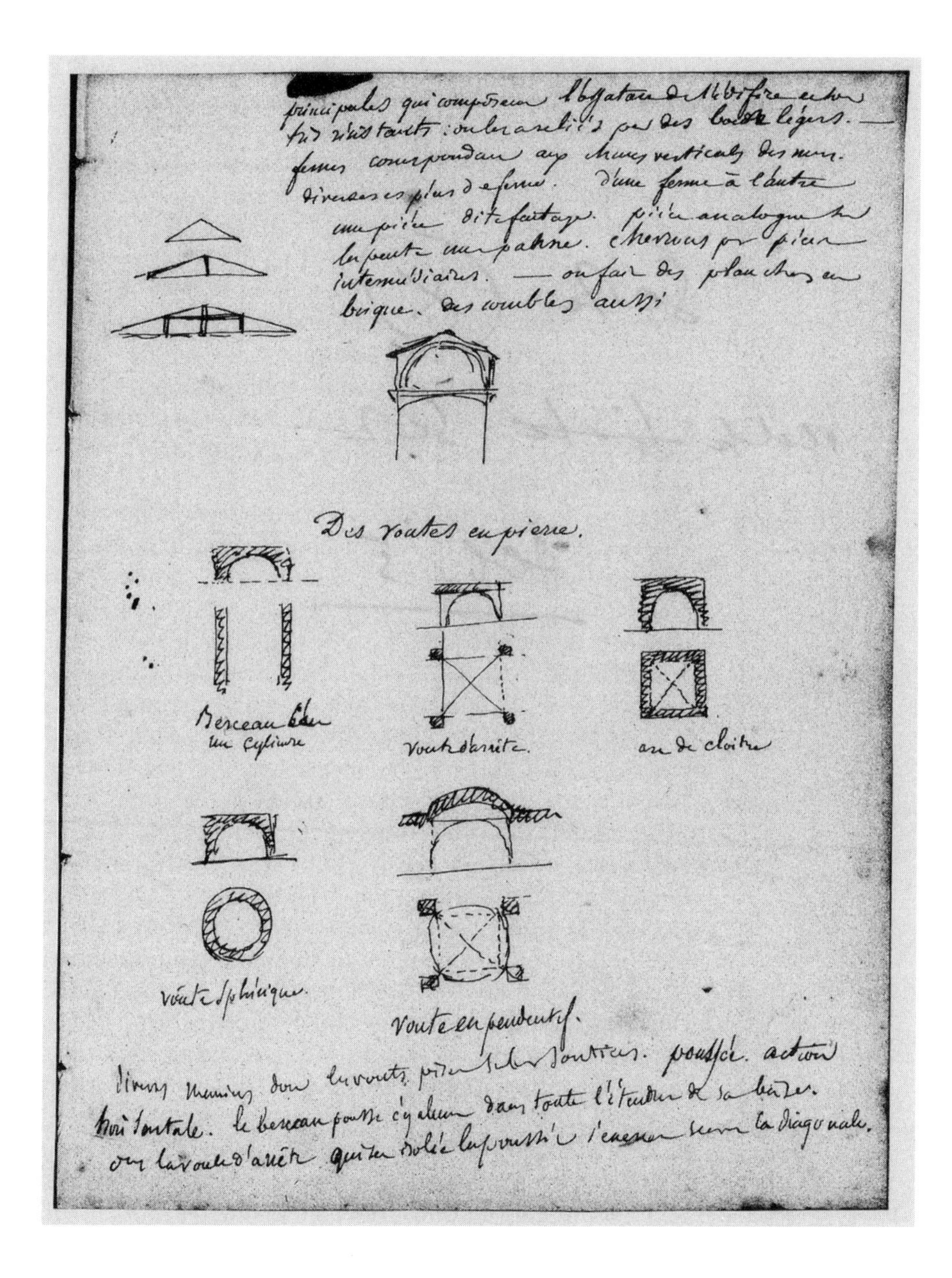

Fransoz illustrates Durand's invention of "building structure": the typological correlation between spatial quality and support-structural order. (Fransoz, 1823)

tailed smaller projects with clear instructions and a detailed assignment that also determined the kind of emphasis to be made in the work.

In Fransoz' notebook one finds also a typology of building structures that show how Durand explained the relationship between rules of space and construction. The various forms of building such as massive, plate, and "skeleton" construction types not only determine spatial quality (qualities of enclosure or directionality, etc.) but in addition also require adequate forms of covering-over: vaults, tunnel vaults, cross-ribbed ceilings, and so on. These basic principles are also to be found in Durand's teaching material "Partie graphique" of 1821.

Didactic material. Durand had to work with a great number of students in a very short period of study. He reacted to this situation by compiling a standardized and efficient methodology of design, which conveyed the basic principles of architecture, planning and design and facilitated the professional realization of building commissions – from stairways to urban planning – as well as the ability to apply the contextual, spatial and building rules that had been learned and practiced in school to unfamiliar and novel assignments.

Included in his didactic working materials and instruments of design (as of 1820) was a standardized form of gridded drawing paper. It measured 45 x 29.2 cm and upon it a square structural grid of 4 cm was outlined in red or orange and divided into four equal parts by a dashed line. Student works from the 1830s, Durand's last teaching years, show how rigorously his "Système quadrillage" was used in exercises.[49]

[49] Szambien, 1984, fig. 106 and 113, in the appendix p.262 and 264.

Another new form of learning introduced by Durand involved trips to view building structures in Paris as illustrative examples of construction. It was a matter of his teaching principle to only view completed structures and not building sites as preferred by the teachers of the École Centrale. Examples of design, design specimens and construction models were also used as illustrations. They were stored in the school's own collection of which a portion were exhibited in the gallery or in the drawing studios as a form of instruction material.

Textbooks. Durand published three textbooks. The first appeared as a collection of plates appearing between 1799 and 1801 under the title: "Recueil et Parallèle des édifices de tout genre, anciens et modernes, remarquables par leur beauté, par leur grandeur ou par leur singularité, et dessinés sur une même échelle." Durand's complete program finds its expression in this title. Different structures, considered by Durand to be exemplary, were compiled in a catalog in order to be open to comparison on an equal scale and to be used in the École Polytechnique's studios as illustration material. Durand used approximately 300 published works and the yield of 18th century travelers. The section with plates began with a preface by the architect J.-G. Legrand including an excerpt from his "Histoire général de l'architecture" of 1799.

In his textbook Durand sought to provide students with the opportunity to become easily and inexpensively acquainted with the many exemplary structures and thus enable them to acquire an overview of Western architecture. By classifying and systematically illustrating objects in the same scale, as well as by reducing the objects to their essential characteristics, Durand pursued didactic goals: The material was to be useful and understandable. In a commentary published in the Journal de l'École Polytechnique of 1799, Durand pointed to the simplified and systematic form of illustrating buildings and emphasized that he did not want to correct them but only clarify their true essence.[50] This teaching material was widely disseminated.[51]

The second textbook was published between 1802 and 1805 under the title: "Précis de Leçons d'Architecure données à l'École Polytechnique" in two volumes. They contained all lessons held by Durand since 1799 and represent the fruit of his (still short) teaching experience. In the first-year course the "Précis" were treated as official, and as of 1810 obligatory material; it was the most expensive of its kind in the school. In the second volume Durand developed a historical typology, such as museums, etc., as well as a "practical typology" of functional buildings of modern architecture of the time. Included in this were among others an observatory, a lighthouse, a large butchery and a covered market. The Précis also contain the methodological essence and operative instruments for the procedure of design, which are actually of Durand's

[50] Durand, 1799.

[51] A new edition was published in 1839 in Belgium and in 1842 in Paris; in 1833 it appeared in two languages in Italy and in 1915 in New York in English; French reprints were published in 1909 and 1936, an American edition in 1981 and a German reprint of the French new edition of 1842 was still brought out in 1986 (Nördlingen, Germany); and at the Parisian École des Beaux-Arts the students used the "Recueil" until 1968! Szambien, 1984, p.100.

making; they will be treated further on in this book. New editions of the Précis appeared until 1840. [52]

[52] Coudray, Durand's first German-speaking student, translated this textbook into German and published it in 1831 in Karlsruhe. The Précis was last reproduced, together with the "Partie Graphique" in 1975 (Nördlingen, Germany)

Finally, Durand published his third textbook in 1821 under the title "Partie Graphique des Cours d'Architecture faits à l'École Royale Polytechnique depuis sa Réorganisation". Durand again demonstrated typological forms of construction as abstract "formulas" that were to be applied to exercise projects or act as a source of inspiration to students. The idealized types of construction again served didactic purposes.

Laying the Foundations of Architectural Theory for Education in Architectural Engineering

Durand's ideas concerning architectural theory were developed for use in education and practical application. They were inspired by the ideas of the Enlightenment and reflected the requirements of post-revolutionary society in France. His experiences in the context of the Académie d'Architecure during the Ancien Régime are coupled with the challenges posed by the approaching industrial era. Increasing focus was placed on issues of utility in architecture and of economy in construction. Both aspects thus form the basis of composition, they are no longer subordinate factors of design but must fulfill the demands of function and the rules of structure, observe the nature of the materials as well as cultural traditions, which determine the specific social and individual needs, preferences and comfort requirements.

Durand's theory of "utility" (Utilité) is grounded on two principles: the kind of function a building is to fulfill and the type of structure that best suits the functional demands. "Economy" (Économie) is to guide the design of a geometrically and mathematically practicable compositional scheme, and thus simplify, clarify and regulate the structural order. In theoretically underpinning the concepts of utility and economy, function (site, program, use), space (composition and aspects of volume) and structure (systems, material) they together formed the methodological key to the designing process which was to characterize modern polytechnical architectural education. What was innovative was that a system of convenient typological, compositional and material rules was developed so that the

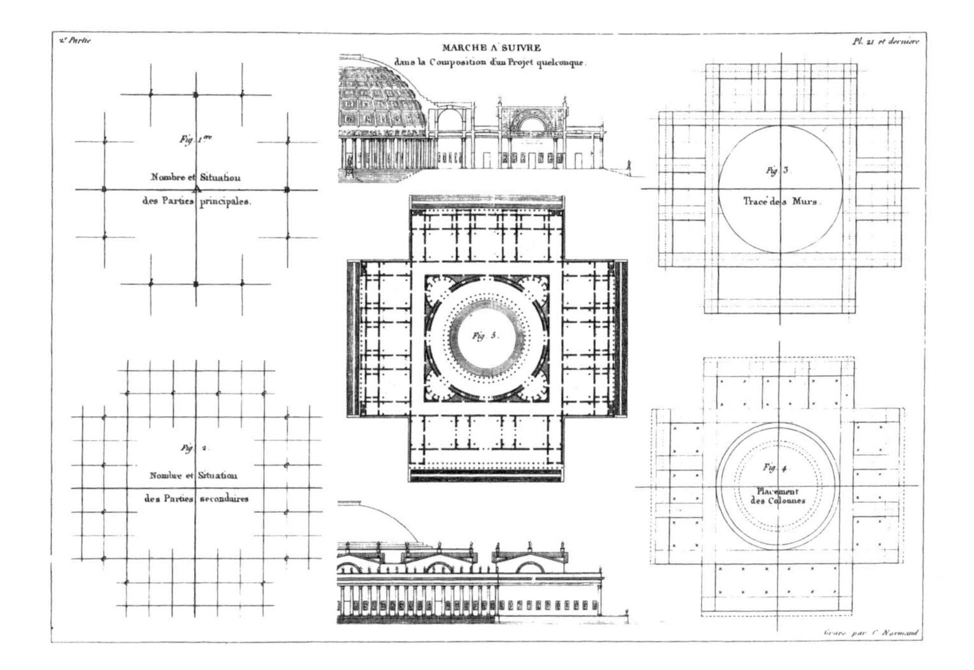

functional requirements of buildings and ensembles could be made more understandable and so that the "economic" structural order derived from it could be based upon formal principles of composition.

The "Marche à suivre" in the "Précis" of 1802–05 illustrates in four steps the basic operations in the design process: basic grid, layout of the room system, wall sections or layers and placement of columns. From this the composition is developed. The spatial order and not the order of columns rules the drawing. (Durand, 1802–1805, 2nd part, Pl. 21)

These theoretical principles originate from a post-revolutionary and already proto-industrial approach to the tasks at hand on the one side, and on the other from a reaction to the 250 lectures that were available and the many numbers of students to be taught. *Durand's architectural theory is indivisible from the didactic demands originating from a limited vocational course of study.*

Didactic aspects. The systematization of the process of designing an architectonic composition is Durand's second innovation. His experience acquired in methodical design while in Boullée's studio was developed further and transferred to his teaching at the École Polytechnique, producing a kind of "major theory of teaching and methodology" that was to remain valid for about forty years. In the form of his Précis he gave his students directions to follow the design process (Marche à suivre) which were further illustrated in the plates.

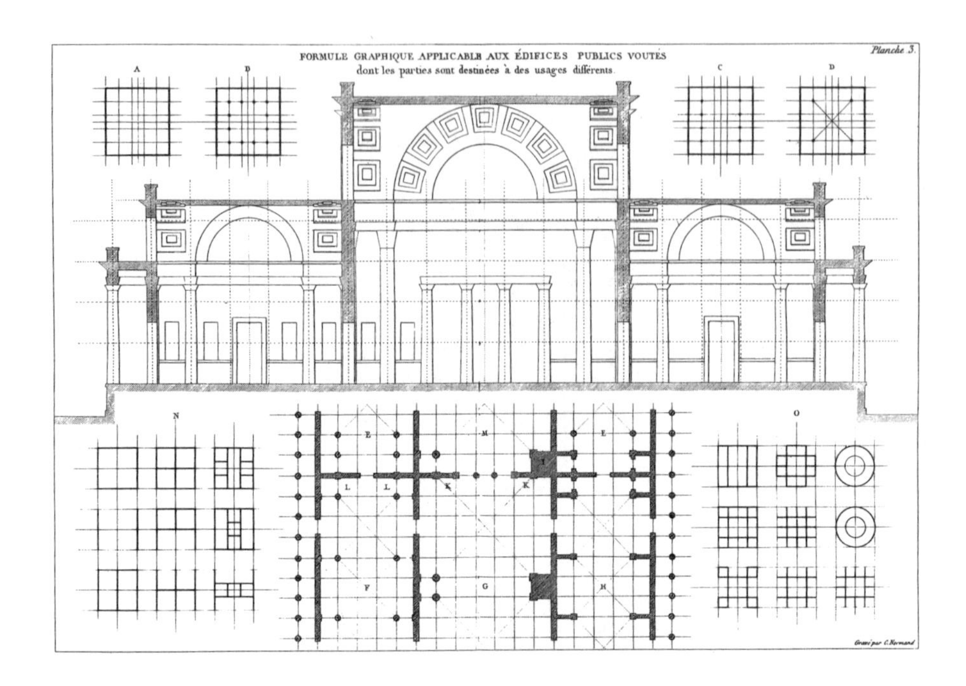

A "Plan Scenario" as drafted on the "Quadrillage" serves to rationally grasp and process construction assignments and allows the assignment to be solved systematically. The grid method is an open system: The structure can be developed from all sides and in three dimensions. (Durand 1821, Pl. 3)

The procedure begins with determining a basic grid; it reflects the specific aspects of the site and delineates the important parts of the building; around it secondary rooms are grouped. As a next step wall-sections are drawn, outlining the wall layers. Ordering the columns occurs as a last step. What is striking here is that the arrangement is made according to the unit-spacing in the axial grid and no longer according to the column-sizes and the distances between them, meaning that the proportions established by the grid dominate the entire arrangement of space, structure and form – this represents an early form of "structural rules of order" (Tracés régulateurs), especially if one takes into consideration a further didactic plate from the Partie Graphique in which working in elevation and ground-plan at the same time is illustrated.

Finally, for the material designing phase Durand included a plate in the Partie Graphique to illustrate the structural elements (Élements des Édifices) of building construction.

Durand defined elements (Unités) as structural units belonging to a specific category or order that de-

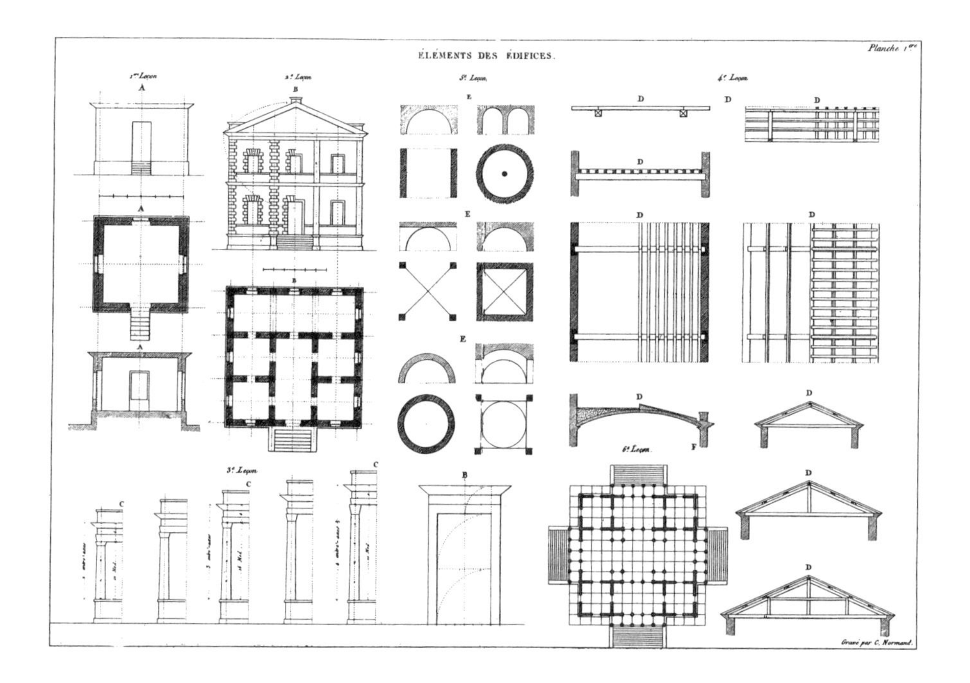

The invention of "building structure": The logical relationship between spatial order and the form and material of structural elements that define it. (Durand, 1821, Pl. 1)

fine space. Together they form a material and formal structure. Basic spatial operations of enclosure (massive construction), directionality (plate or cross-wall construction types), of open forms and non-directionality (column/support or skeleton construction) are defined as the structures of building together with their respective and adequate forms of covering-over (cupola or vaulted ceiling). Durand's typological design specimens also show alternative combination possibilities, as in every ground-plan different spatial demands are to be met.

Durand's conception of "building structure", after Alberti's reduction of walls into skeleton or "ribbed structures" comprised of supporting and "infilling" elements, is a further decisive, theoretical and most importantly methodological shift in paradigm towards the "immaterialization" of the structural system, and paves the way for modern functionalism (cf. "Of the Parts of the Finishing; of Shells, the Stuffing and their Different Sorts", in Alberti, 1755/1986, Book III, Chapter VIII.).

In the Précis Durand ultimately clarified further the kind of materials determined by the building struc-

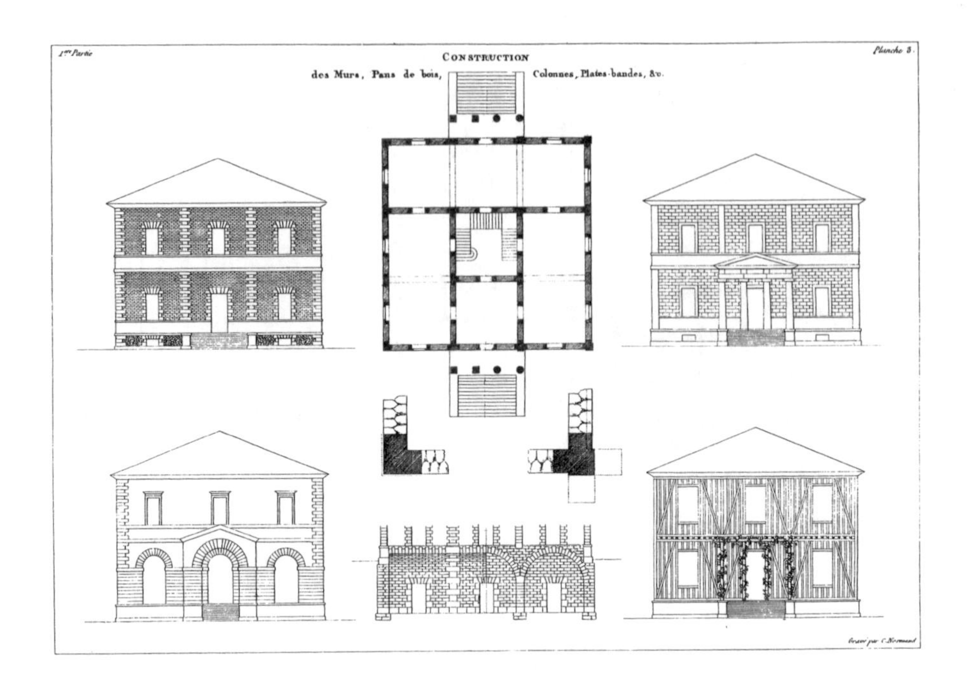

The structural components of exterior walls – piers and fillings – manifest the spatial order of the building's interior. (Durand, 1802–1805, 1. Section, Pl. 3)

ture in a structural table presenting the choice of material for building elements. The building portrayed is made up of a structure of masoned piers. The areas in between are filled with brickwork and dispersed with windows and doors. The non- or self-supporting exterior walls are then surfaced and plastered, however in such a way as to still manifest or visually reinforce the basic order of interior spatial arrangements as prescribed by the structural system – an early form of functionalism! The walls can also be built as wooden post and beam constructions or timber framework. Durand shows four ways (models) to design exterior walls, from a simple house to a Palazzo.

The methodological process. Each step in the process of design is backed up by the didactic tables. According to Durand each task of building construction could be accomplished with it, from the simple stairway to a complex urban installation. Designing and construction become practicable with it. Durand, however, distinguished between the learning process and the teaching method or professional working mode here.

The basic steps in the learning process were organized inductively; meaning an increasing procedure that begins with the most basic elements, then passes on to the structural units until the entire composition can be taken into consideration. In contrast, the teaching method and the last phase in the course of study where the transfer work is done, as well as professional practice itself, entail the deductive working method: the step-by-step process of reducing the general project program or composition first to its structure and finally to its most elemental parts. This method of working, however, requires previous experience.

The Partie graphique teaching materials published in 1821 were to demonstrate how to "decompose" from the ensemble back to the basic elements involved. *In his principles of architectural engineering Durand joined a deductive method of teaching with an inductive order of learning into a complementary and synthetic procedure, in a sense developing further the "Cartesian process of knowledge" into an "architectonic process of knowledge".*

Yet Durand did not consider his standardized method of design as a "recipe book" and the results did not bring forth a "Durand style". *Durand's lessons in architecture were not lessons in style, but lessons in method.* As will be shown further on, his students – such as L eo von Klenze, Charles Rohault de Fleury, Guillaume-Henry Dufour, Charles-Louis Mary – all went very different ways in their professional work.

Durand's Successor: Reynaud and Industrial Orientation

François Léonce Reynaud (1803–1880) studied under Durand in 1821. In 1822 he and Auguste Perdonnet, later the first teacher of railway engineering at the École Centrale, were dismissed from school for their political activity; liberal ideas were not tolerated in the Restoration Regime. In the following years Reynaud was educated as an architect in the sense of the Beaux-Arts, in 1830 he traveled to Italy and subsequently worked as an architect. After the July Revolution he and Perdonnet were rehabilitated by Louis-Philippe. In 1837 the École Polytechnique engaged him to teach architecture, as successor to Durand, and in 1847 he was raised to the status of professor. He taught at this

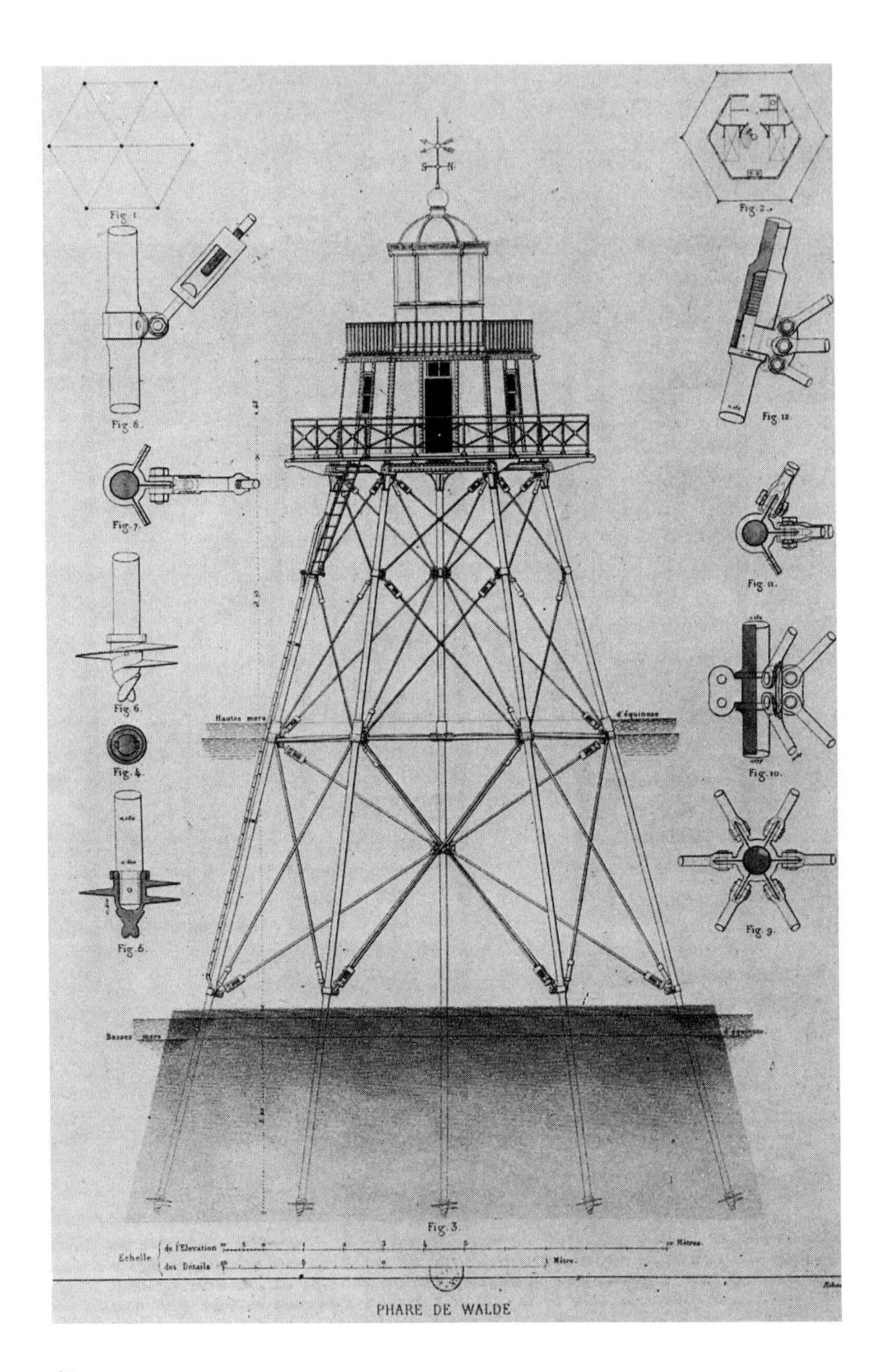
Fig. 1.
Fig. 2.
Fig. 8.
Fig. 12.
Fig. 7.
Fig. 11.
Fig. 6.
Hautes mers
d'équinoxe
Fig. 10.
Fig. 4.
Fig. 9.
Fig. 5.
Basses mers
d'équinoxe
Fig. 3.
Échelle
de l'Élevation
des Détails
Mètres.
1 Mètre.
PHARE DE WALDE

school until 1869. In addition, as of 1842 he also taught architecture at the École des Ponts et Chaussées and became its director from 1869 to 1873.[53] In 1847 Reynaud built the first Gare de Nord station in Paris. His specialty was the construction of light towers and metal structures; between 1846 and 1878 he directed the "Service des phares".[54]

[53] Dartein, in: Reynaud, 1873, p.8.

[54] Reynaud, 1873.

Aspects Concerning Architectural Theory

The methodology and architectural theory underscoring the process of design as established in Durand's principles of architectural education was a decisive factor in the ability of some of Durand's students such as Rouhault de Fleury, Dufour, Mary and others to dare stride the new industrial path and take on the challenges of novel construction assignments, structural systems and materials. Reynaud completed this transition within the polytechnical method of instruction and, in total respect for his teacher and master Durand, surmounted the rigid system of geometry in order to be able to respond to the new demands being made. Durand's principles of utility, of the equivalency of structure and composition, etc. were re-interpreted by Reynaud and transferred into the age of railway engineering. *Under Reynaud, instruction in architecture acquired a new industrial dimension.*

In his contribution to the "Encyclopédie nouvelle" (1836–1841) Reynaud emphasized that the building's form not only follows its function but also the laws that rule the nature of the structural materials and method used.[55] *By adhering to Durand's approach in this way Reynaud accomplished a structural shift in architectural education.* Moreover, according to Picon, in his textbook Reynaud was the first to include information gained from chemical research on the nature of materials, specifically on types of lime, cement and iron, in theory and his teaching.[56]

[55] Reynaud's explanations in the Encyclopédie nouvelle are cited in: Picon, 1992, p.555f.

[56] Elucidation of "Traité d'Architecture" (Reynaud, 1850 and 1858) in: Picon, 1992, p.556–563.

Aspects of Methodology and Teaching

Equally interesting lecture notes as those written during Durand's course have also survived regarding Reynaud, although in far less comprehensive manner. They were written by a student named Kretz from the class of 1851/52 and show that the work being done essentially per-

opposite page: Léonce Reynaud, structural drawing of the light tower Phare de walde, north of Calais, built in 1859. (In: Szambien, 1984, p. 327)

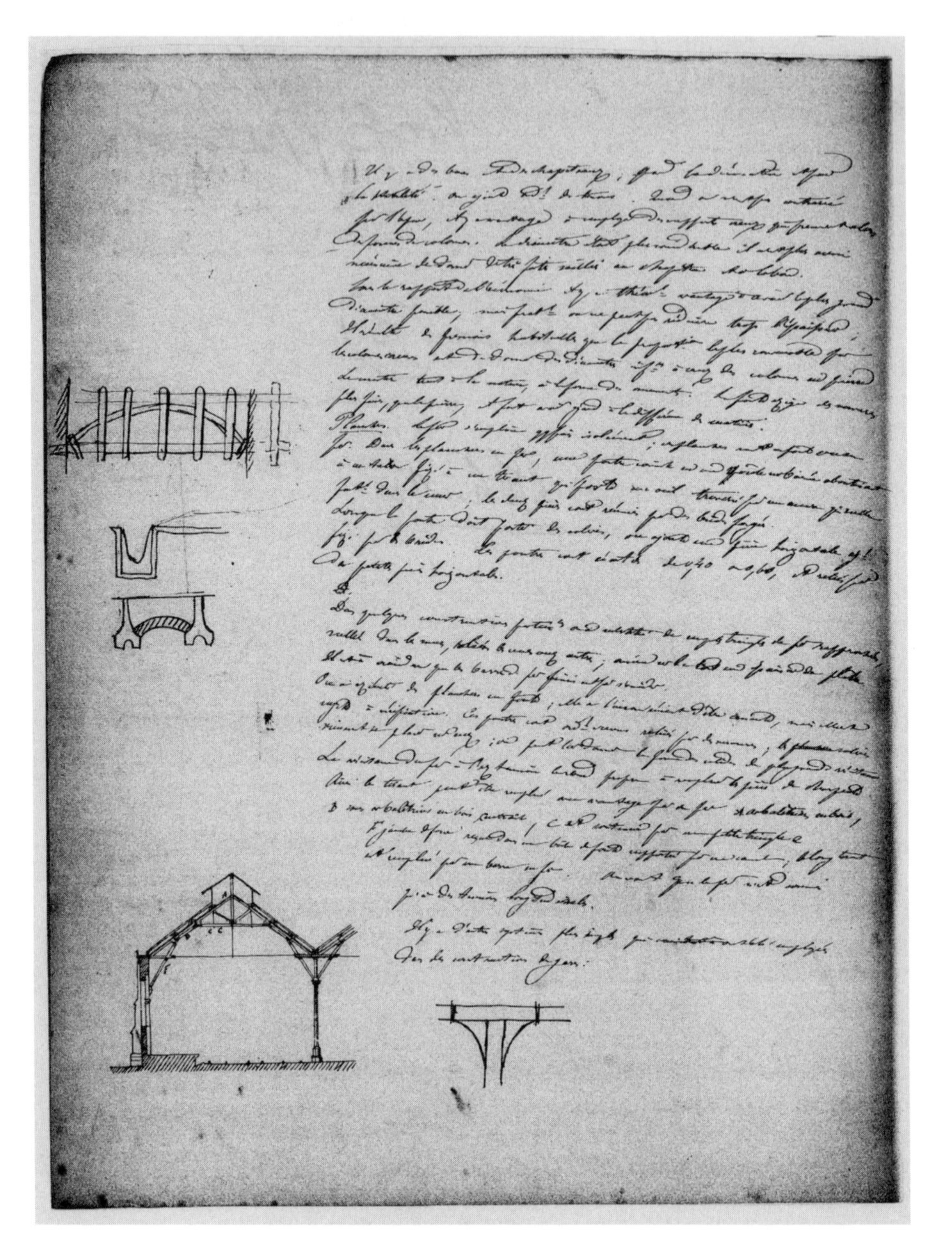

Lecture notes by the student Kretz in Reynaud's course of 1851/52: the modern approach to ceiling and support systems made of iron illustrate the "structural shift" in Reynaud's architecture class. (Kretz 1851/52)

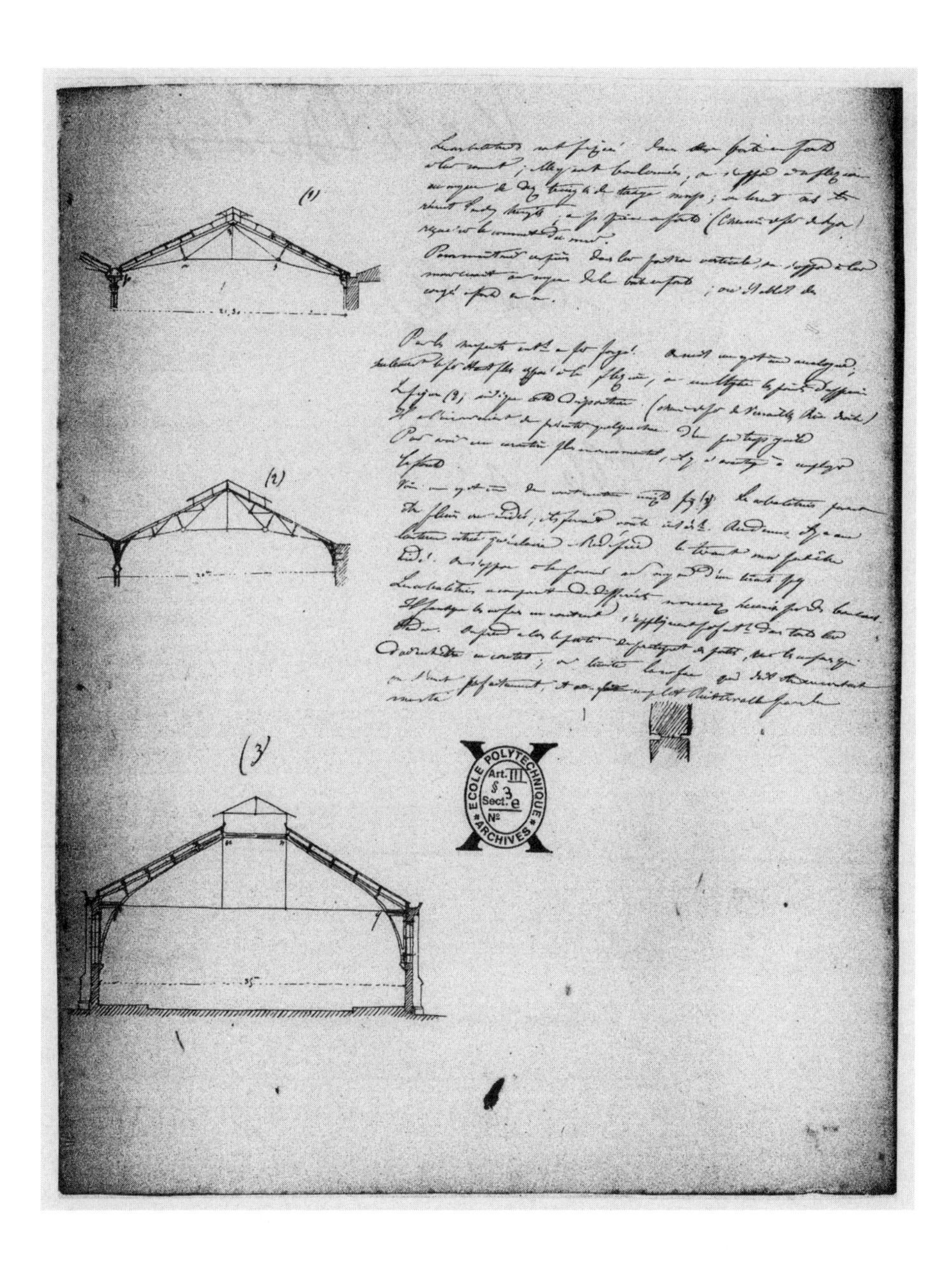
(1)
(2)
(3)
ECOLE POLYTECHNIQUE
ARCHIVES
Art. III
§ 3
Sect. e
Nº

Reynaud's illustrative panel of jointing systems in iron structures: the "state of the art" of architectural engineering in France around 1850. (Reynaud, 1850, Pl. 71)

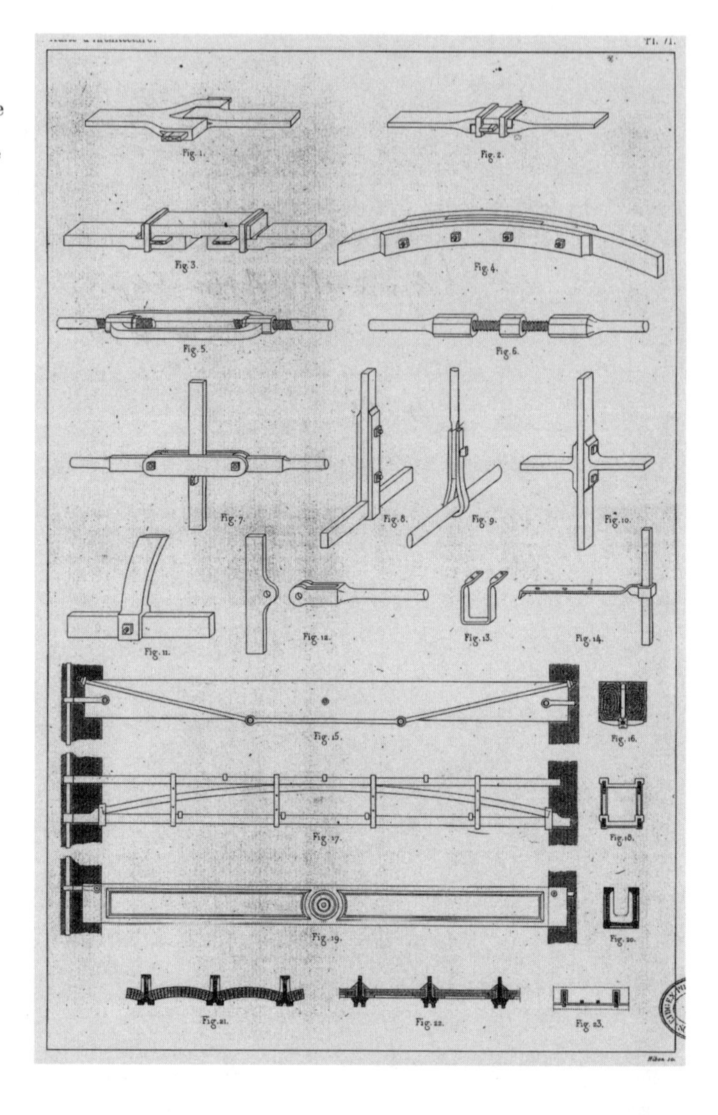

tained to structure and material; they also illustrate Reynaud's industrial orientation in class. Kretz sketched a flooring construction made up of iron supports and a vaulted, rib and tile ceiling as had already been standardized for Central England's cotton mills since 1790. What is to be observed from Kretz's notes is that Reynaud also taught the prestressed roof-frame ("Polonceau trusses")[57] according to Camille Polonceau.[58]

For his course in architecture at the École Polytechnique in the 1850s Reynaud put together a modern text-

[57] Polonceau invented this load-bearing structure in 1837 in the year he accomplished his diploma at the École Centrale, where he was also teacher in the class of 1851/52.

[58] Kretz, 1851/52.

book – at the same time as Mary at the École Centrale – which was to complement the teaching material compiled by Durand. His "Traité d'Architecture" was published in two parts (1850 and 1858) each consisting of a text volume and an illustration volume. [59] Here the most important buildings that had been brought forth by the French industrial revolution were didactically touched upon, such as the Bibliothèque Ste. Geneviève in Paris by Labrouste (which had just been finished in 1850), the Gare du Nord of 1847, planned and designed by Reynaud himself, as well as a few other railway stations.

[59] Reynaud, 1850 and 1858.

The collection begins with an illustration of iron parts assembly ("Assemblages de pièces de fer"), showing the system of iron jointings which had become perfectionized since its first invention in textile industry architecture and later use in English railway stations, as for instance in Turner's Lime Street Station in Liverpool (1849–51).

Thanks to Reynaud, the level of industrially oriented teaching in architecture at the École Polytechnique had reached that of the École Centrale at the middle of the century. Around 1850 both schools took up the new industrial and material challenges and with a similar methodic approach adapted themselves to the modern demands of building. Both teachers, Reynaud and Mary, were students of Durand.

The "Polytechniciens" of the First Hour

Among the graduates of the first two decades since the founding of the École Polytechnique, to be precise the era between 1794 and 1814 and before the Restoration, there are a number of Polytechniciens who stood out as inventors and pioneers of the history of architectural, civil and structural engineering. What is remarkable is that a first group of graduates were to greatly influence French, the second that of German, and the third that of Swiss building history. They all had taken part in Durand's architecture course, a few of them even worked in his studio.

Students from Germany

After the founding of the École Polytechnique, and then after it became known that Durand, one of the most successful architects of the revolutionary era, had become a teacher, many German architects and students traveled to Paris, observed the construction activity – especially those entailing new iron structures such as the Ponts des Arts (1803) – and enrolled in the École or just visited Durand's lectures and his studio. This was to leave its mark on the development of German architecture.[60] Visiting Paris was usually the first stop in a traditional trip to Italy. Friedrich Gilly, teacher of Schinkel and Klenze, traveled to Paris in 1797 and 1798; until 1810 Catel, Karl von Fischer, Friedrich von Gärtner, Harsdorff, Weinbrenner, Hittorff and Gau – later teacher of Semper – and many others followed. Some of Durand's most important students were Coudray, Klenze, Barth and Hess; one of the few students from Berlin was Peter Joseph Lenné, who studied under Durand in 1811 and became one of the most important landscape architects of Europe.[61]

[60] This portrayal follows Szambien, 1984, p.122–133.

[61] On his teaching activity in Berlin, see below.

They did not stay to complete their doctorate but visited the courses and often worked at the same time in Durand's studio instead. Thus also did *Clemens Wenzeslaus Coudray* (1775–1845), who followed the class of 1800/01, drew in his master's studio and was encouraged by him to participate in the Académie nationale d'architecture's competition. In both of the following years he won – as Durand's first German pupil – the Academy's much sought-after "Grand-Prix". Until 1804 he worked as an independent architect in Paris and was also commissioned by Durand to complete the copper etchings for the second part of the Précis de Leçons. Between 1804 and 1808 he stayed in Italy and subsequently returned to Germany, where he began to work as the city architect of Weimar in 1816. In 1831 he completed his translation of the Précis, which he first presented to Goethe, and published as "Abriß der Vorlesungen über Baukunst" ("An Outline of Lectures on Architecture") in Karlsruhe, where a polytechnical school was opened in 1825.

Leo von Klenze (1784–1864), like Schinkel, was trained as an architect under Gilly at the Berlin Academy of Architecture ("Bauakademie") between 1800 and 1803. Gilly imparted the spirit of post-revolutionary, rational and modern Parisian architecture and brought his student's attention to the new systems of iron struc-

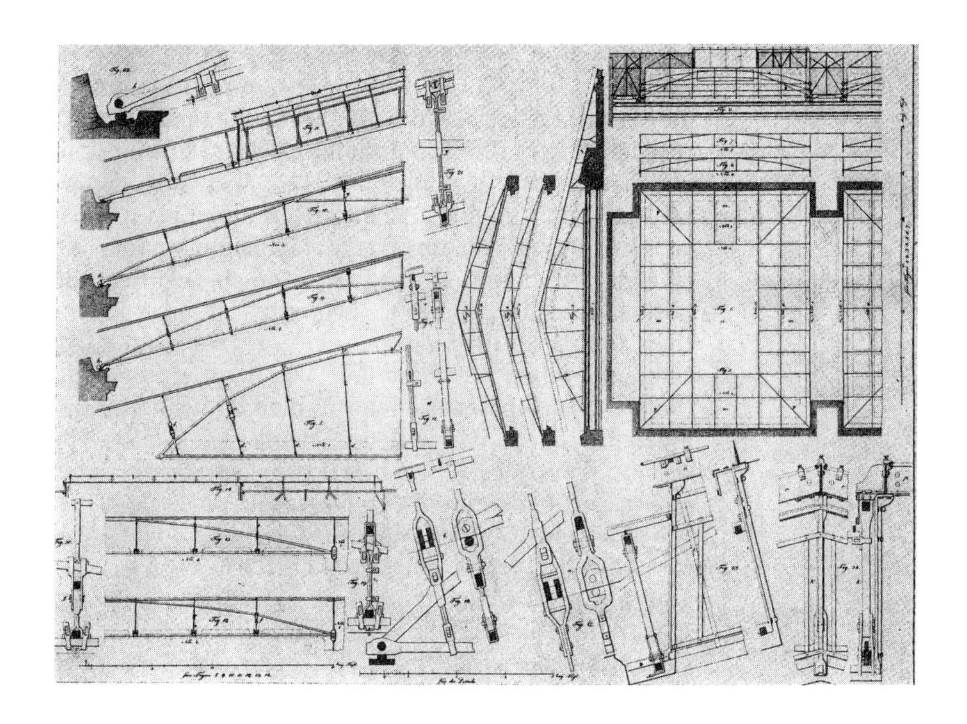

turing. Klenze subsequently traveled to Paris and entered the architectural firm of Percier and Fontaine. Although his presence in Durand's courses at the "École" is not recorded, it has been repeatedly confirmed.[62] As draftsman in Durand's atelier Klenze also used the "Papier quadrillé" and became acquainted with the "Méthode Durand". After 1816 he established himself in Munich and began his great Oeuvre. His Glyptothek can be considered the first German museum to be designed according to Durand's typological principles, and in its spatial arrangement and proportions the Königsplatz refers to the second plate of the second part of the Précis. Among other things, the application of modern French iron structural engineering is to be found in the underpinning of the stage portal in Munich's National Theater, in projects for railway halls and – as a German pioneer achievement – in the structure of Walhalla's iron roof system, which was designed in 1821 and further modified before its construction began in 1830.

In his description of the work content of 1843 Klenze remarked on Walhalla's visible [roof framing structure]: "When finally the use of iron structures

Walhalla's roof structure represents a prototype of the prestressed Polonceau truss. Klenze developed this cast-iron roof-suspension system between 1821 and 1830. It was completed in 1842. The polygonal arrangement of the "suspended posts" inserted between the upper and lower transverse arches are connected by diagonal struts: "The diagonal struts to a great extent lighten the load exerted by the roof and floor structural system onto the principle hanging truss, which then primarily falls on the complete masonry units in the corner." (trans. from Hederer, 1964, p. 154)

[62] Szambien, 1984, p.126.

from England and France were imported to Germany and had become practicable, it became possible to remove the vaulting of the inner hall established to satisfy conditions of non-inflammability."[63] In addition, this structural type allowed the roof area to be generously covered in glass. This for Germany prototypical construction of iron and glass is astonishing when one considers Klenze's other work. Even so, what can be observed is the approach encouraged by his teacher Durand to think in categories of utility and economy in structure, when Klenze writes: "A very sharp demand that the present makes on the art of building is the linking of practical utility with the greatest amount of cost-saving. Working with cast-iron likewise places a new structural element at the architect's disposal."[64]

[63] Trans. from Hederer, 1964, p.309.

[64] Trans. from Hederer, 1964, p.15.

Two other German students were also introduced to Durand by Coudray: *Gottlob Georg Barth* (1777–1848), who studied at the École Polytechnique between 1801 and 1803 and who, after a traveling to Rome, became an influential city architect in Stuttgart as of 1805; and also *Johann Friedrich Christian Hess* (1785–1845), who was later city architect in Frankfurt. He studied under Durand from 1802 to 1803. From this is to be concluded that the influential master builders of the largest and most important cities of Munich, Stuttgart, Frankfurt and Weimar, who were active during the era of burgeoning national unity that was to also find its expression in urban planning, arose from Durand's methodical school of thought and thus carried the polytechnical spirit into their sphere of activity.

Swiss Students

The "Capitulation" agreed upon between Switzerland and Napoleon on 27 September 1803 allowed for twenty students from the Republic of Helvetia a year to attend the École Polytechnique and absolve their training in engineering; one spoke almost of a "Privilège helvétique". As a service in return the Swiss were to provide Napoleon with four permanent regiments of 16,000 soldiers in total to be stationed in Paris. The agreement was dissolved by Louis-Philippe after 1830 and the July Revolution. The long list of regular and external students published by Bissegger illustrate the influence this school had, on Switzerland as well. The first group was made up of Dufour, Pichard and Hegner, who all studied between 1807 and 1809.[65]

[65] Bissegger, 1989, Student list, p.115–151; Bouchet, 1979.

Guillaume-Henri Dufour (1787–1875) came from a family in Geneva. After absolving his polytechnical education (1809) he enrolled in the École du Génie in Metz and received his diploma in 1810 as military engineer. He was subsequently conscripted as a lieutenant in the defense of Corfu against the British, where he stayed until 1814. While there he planned and designed fortifications, compiled topographical maps in which a system of horizontal contour lines was applied for the first time. These he again put to use in his topographical survey of Switzerland ("Dufour Map of Switzerland" 1833–1865). During Napoleon's "Hundred Days" in 1815 Dufour restored the fortifications of the Place de Lyon in Paris, resigned as captain of the French army and returned to Switzerland in 1817.

Guillaume-Henri Dufour as a topographer; portrait by Franz v. Elgger, 1834. (Langendorf, 1987, p.92)

In 1819 he founded a military school in Thun after the French model. In 1827 he was promoted to colonel of the Swiss army, in 1831 he became chief of staff and in 1832–1848 was appointed general of all fortifications in Switzerland. One site that demanded his particular attention in this context was in Saint-Maurice in the Canton of Wallis, where the major link between France and Italy already established by the Romans and further extended by Napoleon over the Simplon Pass could best be controlled. This is where Dufour considered Switzerland's "Pièce de résistance" to be in the defense of its still young commitment to neutrality. In Saint-Maurice Dufour worked with his former fellow students, Hegner and Pichard. In 1847 he was elected General of the later victorious Cantons of the Sonderbund War, which brought Switzerland its national assembly with a federal state and federal constitution.

One of Dufour's lesser known activities pertains to education. Already during his studies at the École Polytechnique his teacher Monge appointed him Répétiteur in the course on Géométrie désriptive; he was also likewise appointed tutor in the drawing course.[66] After his return to Geneva in 1817 he taught Géométrie descriptive in public courses at the Société des Arts in Geneva. In 1819 the Société elected him as their secretary. Until 1822 and then again between 1836 and 1840 he also taught this science of geome-

[66] Chapuisat, 1950.

try and its industrial application in the fields of geodesy and hydraulics at the Académie de Genève, and in 1835 published the corresponding teaching materials comprising 478 pages! Around 1830 (in the same year the École Centrale in Paris was established) he founded an École industrielle at the Société des Arts and worked at the military school in Thun at the same time. In Dufour's teaching activity his scientific intellect and inquiring mind converged with his methodic teaching ability and his leadership qualities.[67]

[67] The historian Mützenberg describes him as a true teacher; Mützenberg, 1991.

Dufour was given the opportunity to become acquainted with exceptional and exemplary teaching personalities during his studies in Paris and Metz. These included Monge, Hachette, Poisson, Sylvestre-François Lacroix (in geometry, mathematics and mechanics), Guyton, Gay-Lussac, Hassenfratz and others (in physics and chemistry), as well as Durand in architecture and city planning. His fellow students also included the later renowned scholars and engineers Poncelet, Fresnel, Coriolis next to Pichard and Hegner from Switzerland – both of them were to become Canton engineers in the Waadt and Zurich each. This scholarly environment and polytechnical school of thought was of decisive importance in Dufour's endeavor to not only maintain personal contact after his return to Switzerland but also to sustain a wide correspondence in scientific circles, to exchange ideas and even to remain on the cutting-edge of scientific, technical and educational developments.[68]

[68] Taton, 1991; Favre, 1947.

Dufour also held important political offices and belonged to the inner circle of founders of the Swiss Polytechnical School in Zurich. As of 1819 he was active as a parliamentarian in the Canton of Geneva for 45 years. In 1841 the Geneva citizens elected him to the Swiss Assemblée constituante fédérale, where he was soon appointed vice-president and where he played an active role until 1847 in the 25–member constitutional committee, which drew up the constitution. In this context and also as member of the new National Council between 1848 and 1851 he worked toward the establishment of Swiss universities and polytechnical schools of higher education, and as of 1851 worked on the Commission on Higher Education.

Since his return to Geneva Dufour also worked as an engineer for the city of Geneva and as of 1827 held

Pont Sainte Antoine, Geneva 1824. (Peters, 1987, p.98; Musée d'art et d'histoire, Geneva)

the office of Chief Engineer of the City and Canton of Geneva. Projects that have since then defined the city, such as the new lay-out of the lake shore and of the Rhone banks in the urban sector, all stem from this period. Under his direction as city engineer the buildings of the botanical gardens and the Musée Rath, among others, were built.[69] In addition, Dufour was active in advancing the railway system, for instance in the construction of the Geneva-Lyon rail line and the central station in Geneva, and was a member of the administrative council of the Swiss Compagnie de l'Ouest.

[69] El-Wakil, 1991.

In 1822/23 Dufour, together with the French engineer Marc Seguin (1786–1875), constructed one of the first and back then largest permanent wire-cable suspension bridges in Europe, the Pont Saint Antoine in Geneva.[70] In the context of this assignment Dufour carried out analyses and material testing with regard to the properties of drawn wire and was the first person to discover their increased strength and greater efficiency compared to suspension bridges of chain construction.[71] These studies were the object of a sci-

[70] In 1822 the brothers Seguin made their first practical experiments with a suspension bridge in Annonay, south of Geneva. In the following according to Peters, 1987.

[71] Straub, 1992, p.235.

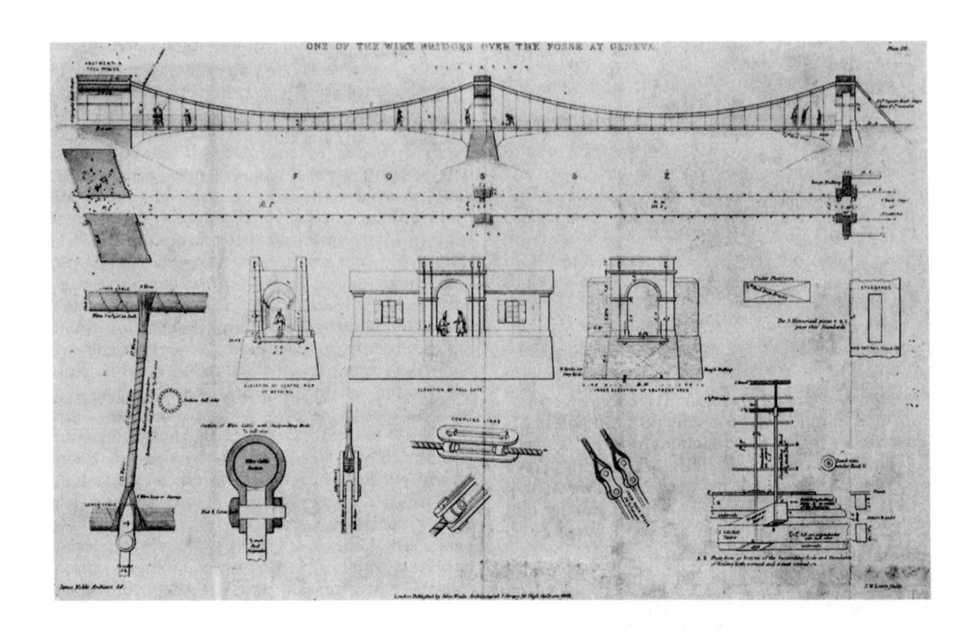

Illustrative Plate of the Pont des Pâquis structure. (Peters, 1987, p. 129; after Weale 1843)

entific treatise he wrote for the Geneva Société de physique et d'histoire naturelle. Dufour built with Seguin a second suspension bridge, the Pont des Pâquis. Both bridges were in the best French Tradition and were influenced by historical reports of English bridge building by Telford, Stephenson and others. Dufour's examination of suspension bridge materials, especially of the application of wire cables instead of chains, is to be included in the context of the work done on the modern theory of elasticity by Navier, professor at the École des Ponts et Chaussées (teacher as of 1819, professor between 1830–1836).

Other Polytechniciens from Switzerland. Salomon Hegner, a fellow student of Dufour at the Paris polytechnical school, built the new Nydegg Bridge in Bern together with another Swiss Polytechnicien, *Ludwig Wurstemberger* (guest student from 1821–1823), a structure that had one massive arch spanning 49,5 meters (1840–1844). Hegner was a bridge inspector of the Canton of Zurich and cooperated with Conrad Escher on the Linth Canal Works in the Canton of Glarus (Linth modification, 1807–1816).[72] Dufour's second student colleague, *Adrien Pichard*, was employed as chief engineer of Ponts et Chaussées and as master builder of the

[72] This work was modeled on the plans made for the Rhine modifications by the engineer Tulla, who founded the Polytechnical School in Karlsruhe; Straub, 1992, p.247f.

Canton Waadt. In this function he built the Grand-Pont in Lausanne between 1839 and 1844.[73]

[73] General Dufour, 1988; Bouchet, 1979.

Before Dufour, Hegner and Pichard attended the École Polytechnique, *Pierre-Joseph Marguet* was a student there (diploma 1803). Together with his son Jules, who studied at the competing school of the École Centrale (graduated 1840), and a further "Centralien", Louis Rivier (1843), Marguet founded the Polytechnical Institute of the University of Lausanne in 1853, the first one of its kind in Switzerland.

Until the Revolution of 1830 a great many Swiss attended the Parisian Polytechnical School and later fulfilled important functions as engineers, scientists, scholars or teachers. One of these was Jean-Samuel Mercanton (1813) who taught at the Academy in Lausanne, Jean-Louis Stapfer (1818) from the Canton of Aargau who was later Switzerland's representative in Paris, and the famous engineer *Auguste Perdonnet* (1821) who was to introduce for the first time in the world courses on railway engineering at the École Centrale. Louis Maurice (1824) was a mathematician and cooperated with Dufour on a program to establish the Swiss Polytechnical School; Henri Ladame (1825) held the chair for chemistry at the Academy of Neuchâtel; Frédérique de Sulzer-Wart (1826) from Winthertur was engaged as Canton inspector for road and bridge construction in Zurich and was also active in the Canton government.

Samuel Vaucher (1798–1877), born in Lausanne, was the only student of Durand from Switzerland to later become an architect. In 1818 he studied further at the École des Beaux-Arts, returned to Switzerland in 1824 and built, among other things, the Musée Rath in Geneva (1824/25) within the context of Dufour's urban planning project. As of 1847 he acted as inspector for the construction of a workers' housing project commissioned by the city of Marseille; in 1861 he built the Palais impérial in Marseille.

Albert de Mousson (1827) studied at the École Polytechnique and at the Sorbonne in Paris at the same time. As of 1833 he taught mathematics at the Zurich Industrial School, gave instruction as private lecturer at the University in that city and subsequently taught experimental physics as a professor between

1855 and 1878 at the newly founded Polytechnikum in Zurich – he was one of its founders and belonged to the first generation faculty.

One of the Swiss students attending the École Polytechnique after 1830 was August Stählin (1832) from Basle, who became a famous railway engineer and joined the board of directors of the Gotthard Railway Company and the Central Swiss Railway. Also belonging to this group was Hypolite Aubert (1833), who acquired a diploma in mathematics at the University of Geneva (1837) and as of 1856 acted as president of the Western Swiss Railway Company. He also worked with Dufour in the military Central School founded in 1819 in Thun. Under Aubert's direction important military – and in today's sense tourist – road projects were realized, such as the Axenstraße along the Vierwaldstätter See, and the Oberalp and Furka passes in Central Switzerland.[74]

[74] Bissegger, 1989.

French Students of the First Generation 1794–1810

Hubert Rohault (1777–1846) was one of Durand's most famous students of the first generation. In 1800 he won second place and in 1802 first place in the Grand Prix of the École des Beaux-Arts. After traveling to Italy he took over important tasks in Paris in public construction after 1806, for instance as building inspector for the Arc de triomphe or as architect for various hospitals between 1817 and 1833 – such as the Hôpital Sainte-Périne, built in 1830 as a garden city arranged within a geometric landscape pattern with individual living quarters and designed according to the rules of composition established by Durand. As of 1819 Rohault became a full-time member of the Conseil des bâtiments civils, his first work consisting of the fish and butter market of 1821 in the area where the later Halles centrales were built and the local constabulary quarters built between 1824–1827 under the direction of his son Charles Rohault de Fleury.

Other Polytechniciens and students of Durand became some of the most important city architects of the time, for instance Auguste-Jean-Marie Guénepin in Saint-Denis, Charles François Ledru in Clermont-Ferrand or Jean Villot in Strasbourg.

Among the pioneers of engineering *Jean Victor Poncelet* (1788–1867) plays a major role. He graduated from the school in 1810 and further developed Monge's geometric approach in mathematics. This allowed novel problems of structural statics to be recorded, treated and demonstrated. Poncelet's invention of a "geometry of properties" of static systems laid the important groundwork for the formulation of "graphic statics" as developed later by mathematicians such as Carl Culmann.[75]

[75] Straub, 1992, p.258f.

The Second Generation of Students 1811–1814
The second generation produced a number of exceptional and famous architects trained by Durand, such as Gilbert, Rohault de Fleury, Reynaud, Rondelet and Mary.

Émile-Narcisse-Jacques Gilbert (1793–1875) entered the École Polytechnique in 1811. He was a friend of Labrouste. In 1820 he received the second prize and in 1822 the first prize of the academy. He began his practical work as a junior inspector in the construction of the Arc de triomphe and later as inspector of the work on the psychiatric clinic in Charenton (1833–1869). He built the police prefecture of Paris and held an important position in the Conseil de bâtiments civils.

Charles Rohault de Fleury (1801–1875) attended Durand's courses at the École Polytechnique from 1820 to 1822 and subsequently received training under his father Hubert Rohault, especially while supervising the construction of the police barracks on the rue Mouffetard and as collaborator on the Passage Saumon (1825), which was the meeting point for Paris fashion until 1899 when it was torn down. He subsequently became general inspector of all barracks on the site of Paris. Rohault de Fleury was not only the builder of elegant and modern greenhouses in the Botanical Gardens of Paris, but also a protagonist in the classicist construction activity of Haussmann's Paris; for instance in the construction of the Grand Hôtel on the Place Opéra, which he designed in cooperation with Armand, Hittorff and Pellechet (inaugurated 1862). It included 700 beds; the ground floor was designed with a cast-iron shop-front and a mezzanine. Its supporting structure was made of iron.[76]

[76] Pevsner, 1976, p.188.

Between 1833 and 1854 Rohault de Fleury built a number of annexed buildings for the Musée d'histoire naturelle, as for instance his famous greenhouse in

The greenhouse built by Rohault de Fleury in 1833 in the Jardin des Plantes in Paris; general view. (Rohault de Fleury, 1867)

the Jardin des Plantes and the addition to it. Besides its iron and glass structure, what is remarkable here is the systematic composition of the entire facility and the typological plan of the central pavilion. It measures 16 x 20 meters and is divided into modules that follow a structural coordinate system of 4 x 4 meters – a direct transference of the "Papier quadrillé" developed by Durand as the basis of the plan. Durand's theory of "utility" and "economy of construction" finds its first modern and spectacular expression in the person of his student, Rohault de Fleury.[77]

In his greenhouse of 1833 Rohault de Fleury achieved the highest level of development possible at that time in the construction of the self-supporting glass facade, which represented the first standardized structure made up of a two-storied, cast-iron supporting frame and puttied glass units. The supports were made up of elements that were screwed together; the middle pieces allowed for joints at the side, i.e. for the connecting visitor's gallery that was constructed along the interior facade at half height. Rohault's pavilion was built within the context of the glass house engineering fashion that was in vogue at that time. It can be said that he took up the experiments undertaken by the Englishman J.C. Loudon concerning greenhouse structure and hot-

[77] In Durands work at least two plates can be found that illustrate structures of modern iron and glass technology, one of which is the "Iron Bridge" in Coalbrookdale of 1779 (Recueil et Parallèle, Pl.23, Paris 1799–1801) and the greenhouse in Liège, Belgium (Recueil et parallèle, Pl.100, Paris 1799–1801)

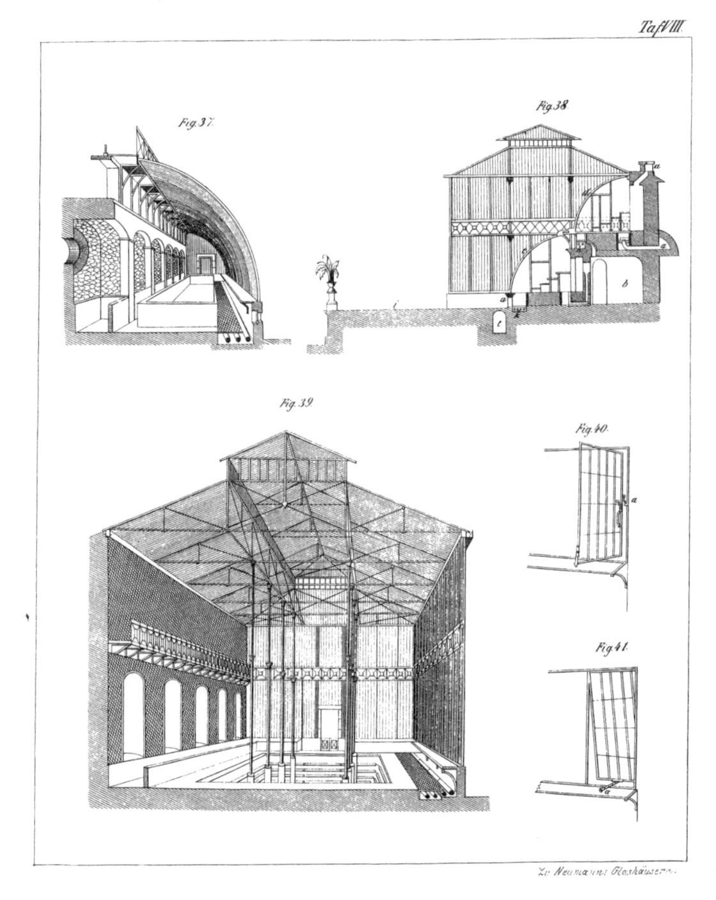

The wrought-iron framework is reinforced by tie members – the precursor to the "Polonceau Truss"; the girders along its length form the supports for the roof-light rods. (Neumann, 1852, Plate VIII.)

house technology, perfected and further developed it, thus paving the way for later structures such as those by Turner or Paxton.[78] Rohault visited England in the same year the greenhouse was built in 1833; his notes recorded during his visit appeared as a series in the Viennese "Allgemeine Bauzeitung" in the year 1837.[79]

A second and further exceptional achievement of his is to be seen in the trussed roof girder, which Polonceau as a graduate of the École Centrale developed as a standardized system of structure in 1837, the year of his diploma ("Polonceau truss"); it was subsequently perfected, e.g. by Turner with his Lime Street Station in Liverpool (1849–51), and used by many other engineers in railway stations.[80] It concerns the very type of spatial supporting structure made of wrought-iron used in Rohault's greenhouse. [color plate 5]

[78] Paxton visited Paris more than once, such as in 1834, and viewed the Pavilion built by Rohault de Fleury in the Jardin des Plantes. Chadwick, 1961, p.93, fn.32.

[79] "Allgemeine Bauzeitung" (Ed. L. Förster), No.48, p.395–400 and No.49, p.403–408, Vienna 1837; in No.32, p.264–266 and No.33, p.271–274 (Vienna 1837), the Rohault de Fleury's glass houses themselves are described.

[80] A more detailed description of Rohault's greenhouses in: Kohlmaier, von Sartory, 1988, esp. p.460ff.

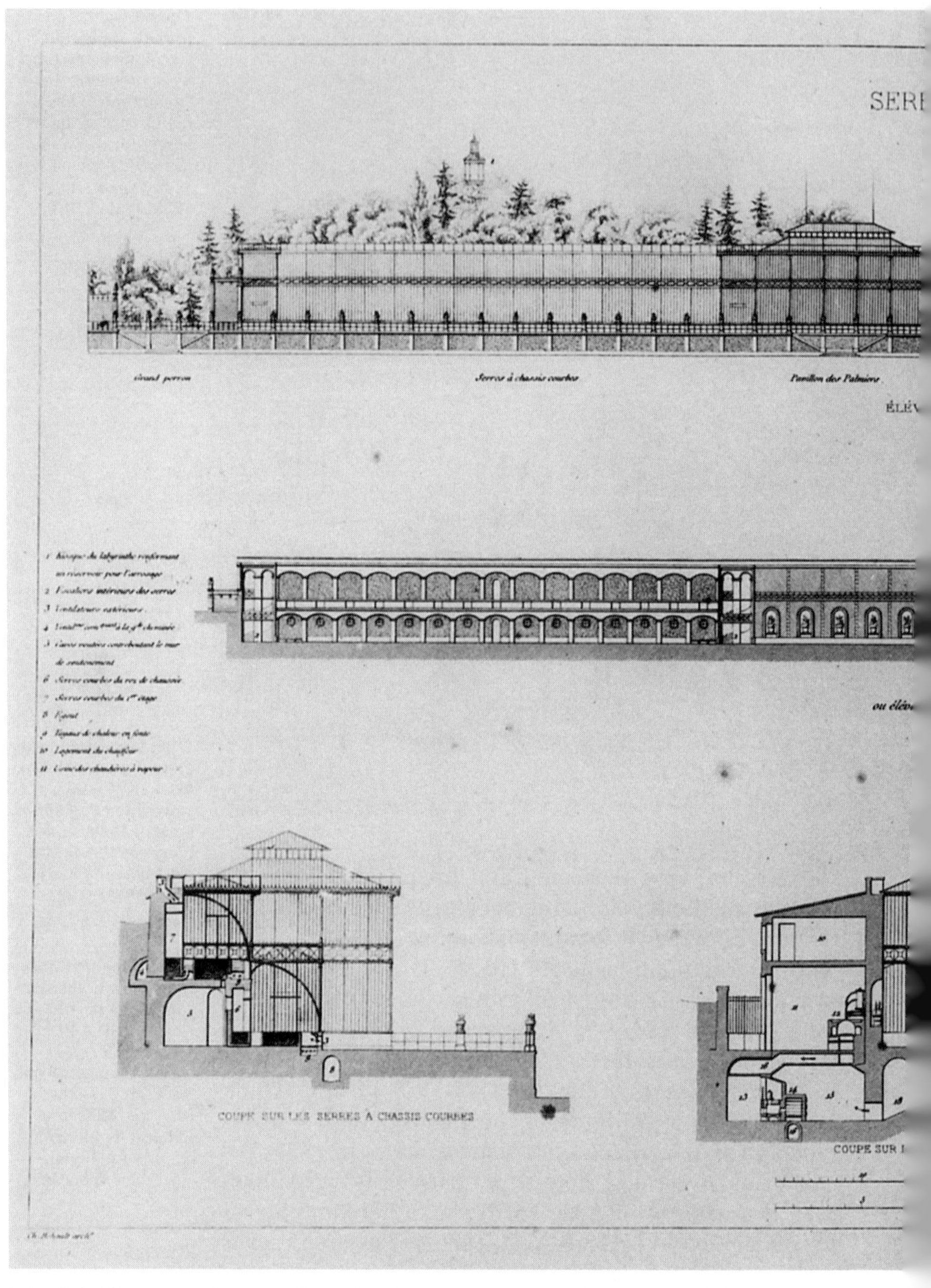

Rohault de Fleury: Greenhouse in the Jardins des Plantes (Paris 1833): sections and elevations. (Rohault de Fleury, 1867)

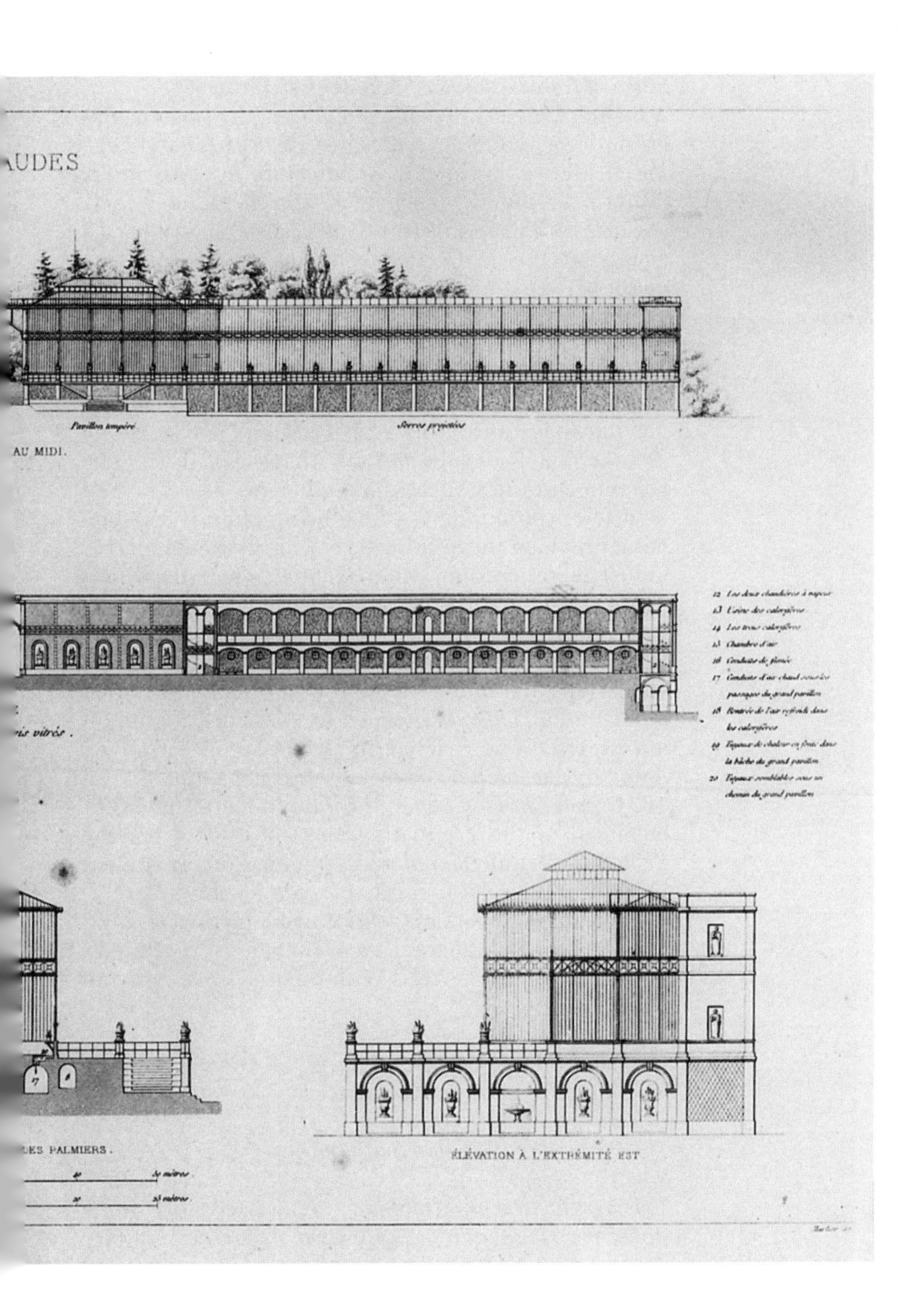
AUDES
Pavillon tempéré
Serres pyriétées
AU MIDI.
ris vitrés.
LES PALMIERS.
ÉLÉVATION À L'EXTRÉMITÉ EST

Some of Durand's other Students from France

A further Polytechnicien belonging to the second generation was *Antoine-Jean-Baptiste Rondelet* (1785–1863). He was trained in architecture not only in his father's studio, a teacher at the École des Arts (Académie d'Architecture), but also in Durand's course at the École Polytechnique. After Durand's death he wrote an obituary and became his most important biographer.[81]

[81] Rondelet, 1835; Szambien, 1984, p.15.

Also to be mentioned are former students of Durand who decisively influenced later teaching and methodology. *François Léonce Reynaud* studied under Durand at the École as of 1821 and subsequently attended the École des Ponts et Chaussées. In 1837 he became Durand's successor – after M. Gauthier had held the position for the time being after 1834. Reynaud provided the architecture course with its further industrial dimension. Other graduates took the initiative and founded the industrially oriented École Centrale at the end of the 1820s, a rival institute of the École Polytechnique and the polytechnical model of education. Its founders and first faculty included among other *Théodore Olivier* (mathematician, diploma in 1811 and teacher in Géometrie descriptive), *Auguste Perdonnet* (railway engineer, diploma in 1821) and *Charles-Louis Mary* (architect). The latter belonged to the graduating class of 1808 from the École Polytechnique and was therefor a fellow student of Dufour, Poncelet and others. As a teacher of architecture at the École Centrale he had a formative influence on an entire generation of "ingenious" architects such as Polonceau, Trélat, Eiffel, Jenney, Moisant and Contamin.

Placing the École Polytechnique in the Context of the Era of Industrialization in France

The Professionalism and Self-Concept of the Polytechniciens

The Polytechniciens took on important cadre functions in post-revolutionary France as civil or military engineers and also as engineers in the administration of the new Republican government. In this regard Monge, for example, was minister for the navy under

Danton; the mathematician Laplace, who was active in the Académie des Sciences, was appointed as minister of the interior by Napoleon in 1799, which oversaw the École Polytechnique; and Arago, a Polytechnicien (1803–1805) and later professor at the "École" (1810–1830), was made minister of the navy in the Second Republic after the 1848 Revolution. Already under Napoleon I, he took on an important role for the school.

A day after his coronation (3 December 1804) Napoleon presents, the regimental sergeant-major Arago with the flag of the "École". (Pinet, 1887, p.52)

While before the French Revolution the royal commission elected members to the committees or councils according to their estate status and background, the educational background of candidates began to play a more decisive role. What was important was the name of the school and the respective influence its corps exerted. This is how the accusation of political "meritocracy" came about, which came to ignore the true quality of education, of professional ethics and the scientific and technical achievements attained by teachers and their students.

Former foreign students of the Parisian Polytechnical School, such as Klenze who carried out diplomatic missions to Greece for the Bavarian king Ludwig I., or Dufour who was a government council for decades in Geneva, general in the Sonderbund War and as of 1848 member of the new Swiss National Council, demonstrate the responsibility that Polytechniciens took on as a matter of course in the context of the political and cultural activity of their new sovereign states.

Often the École Polytechnique is compared to a military cadre school. One of the signs that existed for this was, for instance, the presence of an officer in the local council halls of every one of the 22 provincial cities in which the entrance exams were held in the first years of the École Polytechnique. On the one hand his presence was to represent the military dimension of polytechnical education in post-revolutionary France and remind each candidate at the very beginning of his studies of the ultimate goal involved, and on the other hand to symbolize the new Republic's national unity with its undeniable center of Paris.

Nevertheless, the school's civil aspect must also be taken into account: consider for instance the numerous graduates who went on to specialize at the École des Ponts et Chaussées or at other institutions, even universities. Moreover, a few examples have already been given of Polytechniciens who in the first decades after the school's founding worked as civil engineers and architects in the non-governmental sector, i.e. Rohault de Fleury and Gilbert, or became famous scholars, i.e. Poncelet and Mousson, or who became formative teachers such as Reynaud and Mary.

The polytechnical esprit de corps also finds its expression in the industrial activity exhibited by Polytechniciens and their contribution to the history of science and technology. It often occurred, therefor, that the continuity of an enterprise or the management of an industrial company was less influenced by family tradition than by a "generational succession" of polytechnical graduates. Attending the "École" was and still is a matter of pride, a background one proudly admits to. It is an important and reliable reference and acts as a stamp of quality. Moreover, French families were proud to have their sons trained as Polytechniciens. Similar applies to the École Centrale and the other Grandes Écoles.

However, a certain shift in its understanding as a profession is to be observed, reflecting the changes in the economic and industrial structure. While in the beginning a Polytechnicien was a kind of generalist, scholar as also inventor, architect and engineer as well as officer, trained in the basics of various fields of architecture and engineering, with the advent of industrialization after the 1830s he began to develop into an enterprising dynamic type and finally, in the process of the second industrial revolution and the establishment of large government industrial companies (i.e. as a consequence of the discovery of electricity and fossil fuels, etc.) he developed into a manager responsible for leading major anonymous companies.

On school life. The school life at the École Polytechnique was (and still is) imbued with an esprit de corps that in a sense entailed the acknowledgment of a purposeful common fate. This was formed by the learning communities established in the classes or brigades. The students' daily schedule – in the first phase of the "École" treated here – was rigid and also greatly determined how free-time was spent. From a letter written by the student Bosquet to his mother from the year 1829 we are informed that the (boarding) students were awakened by drums in the school yard at 6 o'clock in the morning: "After 5 minutes one had to get up and immediately go to the drawing studio, where we worked until 8 o'clock." Breakfast followed and at 8:30 the first lecture took place. The breaks were also announced by drums. The lectures lasted until 2:30 p.m. Then came lunch and a break until 5 p.m. Between 5 and 9 p.m. work was done for a second time in the studio or laboratory.

Dinner was held at 9 p.m., after that one returned to one's rooms, went to bed and turned off the lights by 10 p.m., which was again announced with drums; "in the case someone forgot to turn off the light by the last drum stroke, Sergeant would come by – which was not uninteresting to see him for once in his nightshirt instead of in his uniform...."[82]

[82] Trans. from Bosquet (1829), in: Pinet, 1887, Annexe, Note n°14, p.460f.

An identification with the school did not only originate in its disciplinary regimentation of school life or from the career opportunities that the school offered. Rather, it came about especially by the novel teaching model which established the framework for teaching and learning, and which also meant that the teaching individuals in the drawing classes, the Répétiteurs, were knowledgeable and interpersonal role models, that older students or graduates who enjoyed a closer teacher-student relationship to the professors could thus pass on their authority and exemplary role to the students below them. In this way a positive impression of the masters' character was transmitted along a line of increasing hierarchy. This spirit was also carried on proudly by graduates in their work, in their political fields of activity and into the world. Some became Répétiteurs or members of the school council, thus maintaining contact with the school.

In 1865 some of the students and graduates called into life an association of former students, the Association des anciens élèves that was to facilitate the transition from school to practice and to help poorer students. The association founded on 29 November 1866 was the result of the spirit of solidarity and camaraderie that is an expression of the "Esprit de corps polytechnicien".[83] Thus, an identification with one's

[83] Pinet, 1887, p.330f.

PREMI								
JOURS DE LA SEMAINE.	De 5 1/2 à 6 heures.	De 6 à 7 heures.	De 7 à 8 heures.	De 8 à 9 heures.		De 9 à 10 heures.	De 10 à 11 heures.	De 11 heures à midi.
Lundi.	Lever. — Appel. — Prière commune.	Études libres dans les salles.		Déjeûner.	*Jusqu'au 12 Janvier.* Leçon de Physique. *Depuis le 19 Janvier.* Leçon de Géodésie ou d'Analyse appliquée, &c.		Étude sur les matières de la Leçon précédente.	
Mardi.					Leçon et Interrogation d'Analyse ou de Mécanique.		Étude sur les matières de la Leçon précédente.	
Mercredi.					Leçon d'Architecture.	Étude et Travaux sur les matières de la Leçon précédente.		
Jeudi.					Leçon et Interrogation d'Analyse ou de Mécanique.		Étude sur les matières de la Leçon précédente.	
Vendredi.					Leçon de Chimie.		Étude de Chimie. Manipulati[on]	
Samedi.					Leçon et Interrogation d'Analyse ou de Mécanique.		Étude sur les matières de la Leçon précédente.	
Dimanche.		Étude de Belles-lettres, &c.	Déjeûner.	Soins de propreté.	Messe et Instruction, qui seront suivies, selon d'une promenade commune ou de récré			

A normal week's study schedule at the École Polytechnique. (Szambien, 1984, p. 258)

…ISION. *année 1818 –*

… heures.	De 2 à 3 heures.	De 3 à 4 heures.	De 4 à 5 heures.	De 5 à 6 heures.	De 6 à 7 heures.	De 7 à 8 heures.	De 8 à 9 heures.	De 9 à 10 heures.
…e Mécanique ; …, Interrogation …e Professeur, …et demie.	Dîner.	Récréations, Arts d'agrément, Bibliothèque.		Étude de Belles-lettres, &c.	Leçons de Belles-lettres, &c.	Étude de Belles-lettres, &c.	Souper. — Récréations.	Appel. — Prière commune. — Coucher. Lumières éteintes à dix heures.
…omie …ique.				Dessin de la figure et du paysage.	Étude et Interrogation d'Analyse ou de Mécanique.			
…'Analyse …canique, …ogation …phithéâtre, …e cabinet.			Promenade commune et sortie libre, alternativement. En cas de mauvais temps les jours de promenade commune, récréation jusqu'à 4 heures 1/2, et ensuite Études libres.		Étude et de	Interrogation Physique.		
Étude …e Physique ou Machines.				*Jusqu'au 17 Janvier.* Étude d'Analyse. — *Du 17 Janvier au 1.er Avril.* Étude et Interrogation de Géodésie. — *Depuis le 1.er Avril.* Dessin Topographique.	Étude et de	Interrogation Chimie.		
…Analyse …canique.				Dessin de la figure et du paysage.	Étude et Interrogation d'Analyse ou de Mécanique.			
…Mars, …Géodésie. *…Géodésie,* …ique.				*Jusqu'au 1.er Avril.* Étude d'Analyse. — *Depuis le 1.er Avril.* Dessin graphique.	*Jusqu'au 16 Janvier.* Étude et Interrogation de Physique dans le cabinet. — *Depuis le 23 Janvier.* Étude et Interr.on de Géodésie ou d'Analyse appliquée, &c.			
…s.	Dîner et sortie libre. *Comme à la 2e …*				Etudes libres.	Souper et récréations.		

ARCHIVES NATIONALES

school – thanks to which a career and social position was made possible – along with the gratitude felt towards former teachers who one came to appreciate and with whom one often stayed in close scientific and personal contact as well as vital experiences (not only positive ones) were all to become the formative characteristics of the Polytechnicien personality. Proof of this, with respect to Monge and Durand, were already given as examples above.

A personal identification with one's school background and the association established by former students ultimately resulted in a remarkable solidarity with the school, which often was activated to defend against attacks aimed at its very foundations. Critical statements such as those by Olivier mentioned above, for instance, were of a constructive nature. Particularly Olivier, who did not shy from harshly criticizing his school and its development after 1816 and who belonged to the founding members of the industrial rival school, the École Centrale, maintained contact with the school as Répétiteur between 1830 and 1844.[84]

[84] Olivier on the École Polytechnique, in: id., 1851, Preface.

Pioneers of the Industrial Transition

Many other nationalities besides French, German and Swiss students were represented at the École Polytechnique. A list of prominent graduates can be set up for almost every country. A further and complementary perspective arises not only from the history of technology, but also from that of science and of education.

Around the middle of the 19th century the École Polytechnique was once again able to include an industrial orientation in its course of studies. From this time on the Polytechniciens made epochal contributions to the history of science and technology. Nevertheless, even before this turning point the contributions of school graduates are immense and pioneering – which is demonstrated in Arago's review of the practical scientific and technical projects that graduates of the Polytechnical School realized since the school was founded in 1795.[85]

[85] Arago, 1853.

The technological achievements of the first generation. In his paper of 1853 Arago mentions as first examples the international connecting passes over the Simplon, the Mont Cenis and through the Corniche built by the polytechnical engineers Coïc (diploma 1791), Baduel (1797) and Antoine-Rémie Polonceau (1796), as well as the work done by Dufour (1808). According to Arago, the theoretical work on statics by Lamé (1814) and Clapeyron (1816) facilitated the realization of such complex engineering projects as angled bridges or arched railway station roofs, as well as (later) shell vaults in reinforced concrete building.[86] Clapeyron, who completed his studies at the École des Mines (1818), was also engaged in railway engineering (the construction of the first railway line between Paris and St. Germain), constructed the first French steam engine and was a teacher from 1835 at the École des Ponts et Chaussées.[87] On over 25 pages Arago listed up the many projects, procedures, experiments and scientific treatises that demonstrate the pioneering achievements of the first generation of Polytechniciens to have contributed to the history of science, technology and architectural engineering. Nevertheless, the "École" was also outstripped in this respect by the new industrial École Centrale founded in 1829.

[86] Arago, 1853, p.19; cf. also Straub, 1992, p.215, 287 and 293.

[87] Ricken, 1994, p.212f.; Picon, 1992.

On the tradition of polytechnical achievements and inventions up to the present. A general overview that records the school's annals was compiled by Callot.[88] In his paper written in 1982, which has been brought up to date since then[89], the importance of the Polytechniciens in all industrial and infrastructure areas is pointed out.

[88] Callot, 1982, p.365–405.

[89] Lesourne, 1994.

For example, in the field of *public buildings* (besides those graduates already mentioned), Vicat (1804) as founder of the cement industry, Belgrand (1829) who established Paris' water supply system, and Alphand (1836), city architect at the time of Haussmann and construction manager of the World Fair in Paris of 1889 are mentioned; in addition, Mougel (1828), engineer and educated further at the École des Ponts et Chaussées, together with the French engineer Linant worked on the Suez canal project and subsequently held the position as general director of the Suez Canal Company from 1862 until he was succeeded by a further Polytechnicien, Voisin (1838). Furthermore, persons of note active in more recent times in public building are, for instance, the engineer Freyssinet (1879–1962; diploma 1899) who

This page and opposite: Construction site of the first Paris subway; the structure is by Fulgence Bienvenue, "Père du Métro". (Callot, 1982, p.401)

between 1926 and 1929 made the first pioneer experiments with prestressed concrete in bridge building and developed shells and other formations of reinforced concrete (i.e. the airplane hangar in Orly), and finally the specialist for dams, Coyne (1910), who worked on every continent in the world. The frontline status of graduates of the École Polytechnique in the field of engineering is still maintained up into the present.

The names of polytechnical *railway pioneers* are: Legrand (diploma 1809; first national railway network plan), Clapeyron (1816; first passenger line

from Paris to Saint-Germain), Denis (1814; first German railway line from Nuremberg to Fürth), Collignon (1821; senior engineer for the connection between Saint Petersburg to Warsaw and Moscow-Novgorod); further polytechnical engineers worked in Austria (Manuel and Huyot), Spain (Lalanne and Le Chatelier), in Italy (Talabot) and Hunan (Getten). The first organizer of the S.N.C.F. was Le Besnerais (1912). This position also was often held by polytechnical graduates, i.e. Armand (1924) as of 1949 who began with the electrification of the French national railway. Also to be mentioned is Bienvenue (1870), who as an 82–year-old still personally directed the construction of the Parisian subway line No. 9!

Although the Polytechniciens were trained as engineers, they did not always work in this field. An example of this are the philosophers Auguste Comte (1814), Barthélémy Prosper Enfantin (1813) and the social philosopher Georges Sorel (1865), who had a great influence on Lenin and Mussolini. The list of famous Polytechniciens is to be concluded here with the Spanish author Salvador de Madariaga.

Polytechniciens as members of the Institut de France. In taking a look at the list of members of the academic institutes that belong to the Institut de France one can observe that Polytechniciens are strongly represented. As a consequence the history of education and research has been unmistakably influenced by the graduates of this technologically and industrially oriented school type or, polytechnical tradition. In the list published by Callot[90] 16 Polytechniciens (among them Henri Poincaré, diploma in 1873) figure in the Académie Française, 8 in the Académie des Inscriptions et Belles Lettres, 5 in the Académie des Beaux-Arts (including Alphand, 1836), 18 in the Académie des Sciences Morales et Politiques and 154 in the Académie des Sciences (among them Biot, Gay-Lussac, Arago, Malus, Poncelet, Becquerel and many others).

[90] Callot, 1982, p.470–472.

The Origins of a Modern Tradition of Education for Architects and Engineers

Through the ideas developed by the new culture of education the modern architect and engineer attained a similarly respected status in France in the 19th century as that of the scholar in the Ancien Régime. Science and technology were from now on to serve the public welfare, to help improve individual living conditions and not least to contribute to national unity by the all-encompassing distribution of innovative scientific, technological and industrial innovations throughout the new democratic society.

This included making work easier through mechanization (heavy industry, agriculture, etc.), improving the quality of life by introducing standards of health and hygiene (workplace ventilation and lighting, heating and drainage facilities in housing, hospitals and schools), establishing an extensive railway network (opening up markets, improving transportation, social communication), constructing covered markets, galleries and department stores (basic supplies), and so on. Every modern covered market or bridge, no matter in what remote area it stood, every major monument erected in a provincial city – and as of 1830 – every railway line and station, were visible examples of progress and material symbols of modern national sentiment.

As of the middle of the 19th century the national and international exhibitions in Paris increased in popularity and with each time were visited by a record number of people. They not only demonstrated France's leading economic role in the world, in which great pride was taken, but were also opportunities of experimentation for architects and engineers. It is certainly no coincidence that the director of the most important exhibition of 1889, Adolphe Alphand, was a Polytechnicien; his engineers and colleagues were, among others, Eiffel and Contamin, who came from the École Centrale. It was the era of industrial challenge.

With its inauguration in 1794/95 the École Polytechnique initiated a course of development that is still effective today. In quick succession polytechnical institutions were opened in Prague, Vienna and Germany (first in Karlsruhe). At the same time schools were being established according to the Parisian model in America. After 1830 the newly founded École Centrale des Arts et Manufactures introduced a programmatic change and industrially expanded its teaching and learning program.

Sources and References to Further Reading

Alberti, L.B., The Ten Books of Architecture (The 1755 Leoni Edition), Reprint Mineola, New York 1986.

D'Alembert, J.L., Discours préliminaire de l'Encyclopédie (1751), in: E. Köhler (Ed.), Einleitung zur Enzyklopädie, Hamburg 1975 (German/French edition).

'Allgemeine Bauzeitung' (Ed. Ludwig Förster), Wien 1837; Traveling notes of Charles Rohault de Fleury to England, in: N° 48 and 49.

Arago, D.F., Sur l'ancienne École Polytechnique, Paris 1853.

Arago, D.F., Caritat von Condorcet, in: id., Franz Arago's sämtliche Werke (with an introduction by A. v. Humboldt), Vol. 2, Leipzig 1854.

Badinter, E. and R., Condorcet. Un intellectuel en politique, Paris 1989.

Baltard, L.-P., Architecture. Texte de l'Enseignement de cet Art; in: Journal Polytechnique, Paris 1796 [École Polytechnique, Archives: J.E.P., Tome II, Cahier 4, p. 577, 1796].

Belhoste, B., A. Dahan Dalmedico, A. Picon (Ed.), La formation polytechnicienne 1794–1994 (on the occasion of the 200th anniversary of the École Polytechnique), Paris 1994.

Bissegger, P., Étudiants Suisses à l'École Polytechnique de Paris (1798–1850), in: Schweiz. Zeitschrift für Geschichte, Vol. 39, N° 1, Basel 1989, p. 115–151; list of students p. 123–146.

Bouchet, A., L'École Polytechnique et la Suisse pendant la première moitié du XIXe siècle; in: 'Versailles', Revue des amis Suisses de Versailles et de la fondation pour l'histoire des Suisses à l'étranger, N° 647, July 1979.

Chadwick, G.F., The Works of Sir Joseph Paxton 1803–1865, London 1961.

Chapuisat, E., General Dufour 1787–1875, Dietikon 1950.

Callot, J.-P., Histoire de l'École Polytechnique, Paris 1982.

Comberousse, Ch. de, Projet de Rapport à Présenter au Ministre de l'Agriculture et du Commerce, dated probably 1869 [Archives de l'Académie des Sciences, Paris].

Comité de salut public, Program and Courses of the École de Mars, in: M.J. Guillaume, Index: 2140 (INRP), Paris 1948.

Condorcet, La vie de Turgot (1786), in: Œuvres de Condorcet (Ed. A. Condorcet O'Connor and F. Arago), Vol. 5, Paris 1847; Reprint Stuttgart/Bad Cannstatt 1968.

Condorcet, Sur l'instruction publique. Mémoires (1791), in: Œuvres de Condorcet (Ed. A. Condorcet O'Connor and F. Arago), Vol. 7, Paris 1847; Reprint Stuttgart/Bad Cannstatt 1968.

Condorcet, Rapport et Projet de Décret sur l'Organisation générale de l'Instruction publique, Paris 20 and 21 April 1792, in: Œuvres de Condorcet (Ed. A. Condorcet O'Connor and F. Arago), Vol. 7, Paris 1847; Reprint Stuttgart/Bad Cannstatt 1968.

Dhombres, J., L'École polytechnique et ses historiens, Introduction to: A. Fourcy, Histoire de l'École Polytechnique, Paris 1828 (Reprint Paris 1987).

Dhombres, N., Les Savants en Révolution 1789–1799, Exhib. cat. Cité des Sciences et de l'Industrie, Paris 1989.

General Dufour. Der Festungsbauer, Exhib. cat. Kant. Museum Altes Zeughaus, Solothurn 1988.

Dupin, Ch., Essai Historique sur les Services et les Travaux Scientifiques de Gaspard Monge, Paris 1819.

Durand, J.-N.-L., Recueil et Parallèle, in: Journal de l'École Polytechnique, Paris 1799 [École Polytechnique, Archives: J.E.P., Cahier 11, p. 361ff., Paris 1799].

Durand, J.-N.-L., (with J.G. Legrand), Recueil et Parallèle des édifices de tout genre, anciens et modernes, remarquables par leur beauté, par leur grandeur ou par leur singularité, et dessinés sur une même échelle, Paris 1799–1801.

Durand, J.-N.-L., Précis des Leçons d'Architecture données à l'École Polytechnique, 2 Vols., Paris 1802–1805 (New edition: Nouveau Précis, Paris 1819 resp. 1817).
Durand, J.-N.-L., Partie Graphique des Cours d'Architecture faits à l'École Royale Polytechnique depuis sa Réorganisation, Paris 1821.
Durant, W. and A., Kulturgeschichte der Menschheit, Vol. XXXII., Lausanne (original title: The Story of Civilization, 1935).
El-Wakil, L., G.-H. Dufour et le nouveau visage de Genève, in: Durand, R. and D. Aquilon (Ed.), Guillaume-Henri Dufour dans son temps 1787–1875. Actes du colloque Dufour, Genève 1991, p. 215–230, p. 199–214.
Favre, H. (Ed.), L'œuvre scientifique et technique du Général Guillaume-Henri Dufour, Neuchâtel 1947.
Fourcy, A., Histoire de l'École Polytechnique, Paris 1828 (Reprint, in: J. Dhombres, 1987).
Fransoz, A., Lecture Notes from the Course J.-N.-L. Durand 1823 [École Polytechnique, Archives: Titre III, Section 3, Carton N° 2, 1816–1929].
Furet, F. and M. Ozouf (Ed.), Dictionnaire critique de la révolution française, 4 Vols., Paris 1992.
Germann, G., Einführung in die Geschichte der Architekturtheorie, Darmstadt 1980.
Grelon, A., Von den Ingenieuren des Königs zu den Technologen des 21. Jahrhunderts. Die Ausbildung der Ingenieure in Frankreich, in: A. Grelon, H. Stück (Ed.), Ingenieure in Frankreich, 1747–1990, Frankfurt a/M./New York 1994, p. 15ff.
Griewank, K., Die Französische Revolution 1789–1799, Graz-Köln 1958.
Guillaume, M.J. (Ed.), Note sur l'Instruction Publique de 1789 à 1808 suivie du Catalogue des Documents Originaux existant au Musée Pédagogique, Paris 1948; the education concepts and documents listed here are to be found in the archive of the Institut National de la Recherche Pédagogique library (INRP), in Paris under the 'Index Guillaume'.
Guyton, L.B., Projekt zur École de Mars, in: M.J. Guillaume, Index: 11823 (INRP), Paris 1948.
Hachette, J.N.P., Traité de Géométrie Descriptive, comprenant les applications de cette géométrie aux ombres, à la perspective et à la stéréotomie, Paris 1822.
Hager, F.-P., Staat und Erziehung bei Rousseau, Helvétius und Condorcet – ein Vergleich, in: F.-P. Hager, D. Jedan (Ed.), Staat und Erziehung in Aufklärungsphilosophie und Aufklärungszeit, Bochum/Wien 1993, p. 67–95.
Hederer, O., Leo von Klenze. Persönlichkeit und Werk, München 1964.
Julia, D., Les trois couleurs du tableau noir. La Révolution, Paris 1981.
Kohlmaier, G., von Sartory, B., Das Glashaus, ein Bautyp des 19. Jahrhunderts, München 1988.
Kretz, Lecture notes from the Course Reynaud, 1851/52 [École Polytechnique, Archives: Titre III, Section 3, Paragraphe e, Carton N° 1, 1794–1991].
Langendorf, J.-J., Guillaume-Henri Dufour. General – Kartograph – Humanist, Zürich 1987 (French edition, Luzern 1987).
Langins, J., La République avait besoin de savants. Les débuts de l'École polytechnique: l'École centrale des travaux publics et les cours révolutionnaires de l'an III, Paris 1987; incl. among other things reprints of the following documents: 'Programmes de l'Enseignement Polytechnique de l'École Centrale des Travaux Publics', 8 February 1795 (Annexe G, p. 126ff., with 'Architecture', p. 146–162); 'Rapport et Projet de Décret' by Fourcroy, 28 September 1794 (Annexe H, p. 199ff.); 'Développements sur l'Enseignement', September 1794 (Annexe I, p. 226ff.).
Lesourne, J., Les Polytechniciens dans le siècle 1894–1994, Paris 1994 (Bicentenaire de l'École Polytechnique).
Michelet, J., Histoire de la Révolution Française (1847), Éd. Gérard Walter, 2 Vols., Paris 1952.

Le Moël, M, and R. Saint-Paul (Ed.), Le Conservatoire National des Arts et Métiers au coeur de Paris 1794–1994, Paris 1994.
Monge, G., Stéréotomie, in: Journal Polytechnique, ou Bulletin du Travail fait à l'École Centrale des Travaux publics, Paris 1794 [École Polytechnique, Archives: J.E.P., Tome I, Cahier 1, 1794].
Monge, G., Géométrie Descriptive, Paris 1801 (Reprint Sceaux 1989).
Mützenberg, G., Dufour enseignant, in: Durand, R. and D. Aquilon (Ed.), Guillaume-Henri Dufour dans son temps 1787–1875. Actes du colloque Dufour, Genève 1991, p. 77–86.
Napoleon Bonaparte (Ed.), Description de l'Egypte, Paris 1802 (Reprint Köln 1994).
Napoleon, Correspondence to the governor of the École Polytechnique of 23 March 1805 [École Polytechnique, Archives: Titre I, Section 2, Carton N° 1, 1794–1815].
Neumann, M., Grundsätze und Erfahrungen über die Anlegung, Erhaltung und Pflege von Glashäusern aller Art, Weimar 1852 (Reprint Wiesbaden/Berlin 1984).
Olivier, Th., Mémoires de Géométrie Descriptive, Théorique et Appliquée, Paris 1851; incl.: De l'École Polytechnique (Introduction).
Perdonnet, A., Traité des chemins de fer, Tome II, Paris 1860.
Peters, T.F., Transitions in Engineering. Guillaume Henri Dufour and the Early 19th Century Cable Suspension Bridges, Basel 1987.
Pevsner, N., Academies of Art. Past and Present, New York, 1973.
Pevsner, N., A History of Building Types, London 1976.
Picon, A., Architectes et Ingénieurs au Siècle des Lumières, Marseille 1988.
Picon, A., L'invention de l'ingénieur moderne. L'École des Ponts et Chaussées 1747–1851, Paris 1992.
Pinet, G., Histoire de l'École Polytechnique, Paris 1887.
Reynaud, L., Traité d'Architecture, 2 Vols., with a volume of plates each, Paris 1850, resp. 1858.
Reynaud, L., Les Travaux Publics de la France, Paris 1873.
Ricken, H., Der Bauingenieur. Geschichte eines Berufes, Berlin 1994.
Rondelet, J.-B., Notice historique sur la vie et les ouvrages de J.N.L. Durand, in: 'Moniteur', Paris 3 January 1835.
Rohault de Fleury, Ch., Muséum d'histoire naturelle: serres chaudes ecc., Paris 1837 [Bibliothèque Centrale du Muséum National d'Histoire Naturelle, Paris].
Schmitt, E., Einführung in die Geschichte der Französischen Revolution, München 1980.
Shinn, J., L'École Polytechnique 1794–1914, Paris 1980.
Soboul, A., Kurze Geschichte der Französischen Revolution, Berlin 1989.
Steiner, G., Lernverhalten, Lernleistung und Instruktionsmethoden (Ed. Institute of Psychology, University of Basel), Basel 1993; id. also in: F.E. Weinert (Ed.), Enzyklopädie der Psychologie, Vol. II: Psychologie des Lernens und der Instruktion, Göttingen 1993.
Straub, H., Die Geschichte der Bauingenieurkunst. Ein Überblick von der Antike bis zur Neuzeit, Basel 1992.
Szambien, W., Jean-Nicolas-Louis Durand.1760–1834. De l'imitation à la norme, Paris 1984.
Taton, R., L'Œuvre Scientifique de Monge, Paris 1951.
Taton, R., Guillaume-Henri Dufour et la science de son temps, in: Durand, R. and D. Aquilon (Ed.), Guillaume-Henri Dufour dans son temps 1787–1875. Actes du colloque Dufour, Genève 1991, p. 153–168.
Tocqueville, A. de, L'Ancien Régime et la Révolution, Paris 1856.
Villari, S., J.N.L. Durand (1760–1834). Art and Science of Architecture, New York 1990.
Vogt, A.M., Boullées Newton-Denkmal. Sakralbau und Kugelidee (Publication series of the Institute for the History and Theory of Architecture at the ETH Zürich, Vol. 3), Basel 1969.
Vogt, A.M., Boullée – Architektur. Abhandlung über die Kunst (Ed. B. Wyss); Introduction and commentary by A.M. Vogt, Zürich/München 1987.
Voltaire, L'affaire Calas et autres affaires (Ed. J. Van den Heuvel), Paris 1975.

The École Centrale des Arts et Manufactures in Paris

France's strongly progressing industrialization, which began in 1815 after Napoleon's fall, also made great demands on the École Polytechnique it could no longer completely fulfill. The school had become too fixated on its acquired role as a training ground for new cadres intended for government officer and engineering corps. In addition, under the influence of the Restoration regime the "École's" teaching program continued to place greater focus on theoretical subjects, most importantly on mathematics, and to reduce teaching in the "Sciences appliquées".

The Establishment of an Industrial Culture of Education

With the restoration of the monarchy, the otherwise traditionally unlimited access to the initial Polytechnical School as defined by the Enlightenment changed radically after 1815 in that admission was regulated according to the government's need of military and civil engineers and officers, and the age of leave was reduced. As a result, the "École" was now given its orientation as an elite school. The newly established École Centrale des Arts et Manufactures of 1829 (in the following as École Centrale) was thus able to recruit most of its students from the body of candidates not accepted to the École Polytechnique and was therefore able to quickly expand! For example, in 1835, 620 candidates hoped to enroll in the École Polytechnique, but only 150 were accepted. The number left over was much greater than the École Centrale's enrollment capacity.[1]

It is no coincidence that the first group admitted to the new school included former "Polytechniciens". They perceived the signs of the time, demanded a modern and industrially oriented education for architects and engineers and hoped to become qualified in the practice of science, technology and industry. The establishment of the École Centrale in the year 1829 is consequently also the result of the new kind of demands being formulated.

[1] Weiss, 1982, Table 1.1, p.20 (In the period between 1810–1850).

Hence, the polytechnical model as maintained in its first decade (1794/95–1804) served as a guide and as the basis for the new school, especially with regard to the scientific principles of engineering formulated in a standardized curriculum valid for all students and also with regard to the methodological combination of theory and praxis. "La science est une!" (a "sole science") was the guiding principle of the École Centrale's founders; and with it they again made reference to the philosophical ideals of the Enlightenment, particularly to d'Alembert and Condorcet, as did those who initiated the Polytechnical School.[2] However, the École Centrale's model of instruction was substantially expanded in its industrial dimension.

[2] The previous chapter already touched upon the epistemological "Tableau" of knowledge, cognition and perception and its transfer to practice.

Saint-Simon's positive view of the industrial revolution and his almost propagandistic activity had a great impact on those circles who were engaged in forming France's scientific, technological and industrial future. The École Polytechnique had already brought forth its "Philosophes", such as Auguste Comte (diploma 1814) or Barthélémy Enfantin (1813) who founded the circle of "positivism"; yet not until the new industrial school was established could the ideas formulated by this circle be made manifest and have an impact on the goals of education. The founding of the École Centrale was based on this spirit. It was also a time of historical upheaval, of the dissolution of the Restoration initiated by circles carrying political responsibility in this new age of industrialization.

Saint-Simon and the industrial revolution. Saint-Simon lived until 1825. In his autobiographical notes he stresses that after the French Revolution he had sought to establish a large industrial firm and a scientific school of the most perfect quality: "In a word, I wanted to contribute to knowledge and to the betterment of the human fate; this was my true ambition". With regard to scientific and industrial progress Saint-Simon developed his theories of "positivism" and "industrialism". According to Saint-Simon, the philosophy of history is derived from the power of science, or the intellect, which also determine economic and material progress. In his social theory he formulated the term "industrial" to encompass the entire working population engaged in socially useful work. Here Saint-Simon differentiated between "Savants d'application" (producers) and "Industriels de théorie" (scholars). Connected to this is the concept of "utopian socialism", associated with the theories of Owen and Fourier. After Saint-Simon's death and the founding of a new industrial school in 1829, "Saint-Simonism" continued, evolved in different directions and later addressed the consequences of industrialization, of impoverishment, pauperism, etc. From this and within the context of the new workers' movement socialist concepts were developed, which in part were related to the ideas formulated by Marx and Engels. Saint-Simon's concept of the "power of ideas", however, was considered by Marx to be a form of political reformism.[3]

[3] Saint-Simon, 1821 and 1823/24, as well as Salomon, 1919.

The Invention of "Industrial Sciences"

Social and economic practice, as reflected in and having an influence on the new school after 1830, were completely different thirty years after the founding of the École Polytechnique. The nature of building programs had changed, theoretical and methodological principles had been substantially extended. The founders and first generation of professors of the École Centrale created a new system of "industrial sciences" that was to combine basic sciences with nascent industrial engineering practice; from this they developed an industrially oriented model of instruction. The role that the "Géométrie descriptive" played in earlier polytechnical instruction as a means of combining theory and practice was now to be played by the "Sciences industrielles". These "intermediate subjects", as the interface between basic principles and industrial practice, were titled "industrial physics", "industrial chemistry", "mechanical engineering", "industrial, structural and architectural engineering", etc.

The École Centrale: A Model for the Establishment of Schools in Europe and America

The school's modern program for architects and engineers was widely and swiftly accepted by the cutting-edge of France's industrial development: its graduates were very popularly sought-after. In just a short time the École Centrale attained a leading status among higher institutions of scientific and technological education world-wide; it had a great impact on the establishment of schools throughout Europe and the USA and quickly became the focal point for students and teachers from countries all over the world. Just as the École Polytechnique became the universal model for similar new schools around 1800, as of 1830 most of the new scientific and technological colleges of engineering and industry were based on the École Centrale model, which is exemplified by schools in the USA. The Polytechnical School in Karlsruhe even underwent an institutional shift: although it was established according to the polytechnical model in 1825, in 1832 it appropriated the industrial system of instruction introduced by the Parisian École Centrale and finally, in 1841, it developed its own organizational model, which again served as the basis for the Swiss Polytechnical School in Zurich in 1855.

The École Centrale's novel course in railway engineering, and in mechanical and industrial architectural engineering commanded a great deal of interest. The goal was to provide basic training to generations of architects and engineers, which as a form of "Armée industrielle" would allow them to work towards social progress in all countries. Wherever they worked these "Ingénieurs industriels" were considered almost as benefactors in their endeavor to raise living standards, alleviate work and improve hygienic conditions. They built bridges and railways, covered markets and greenhouses, heated homes and hospitals, ventilated and well-lighted factories; they improved factory security and built leisure-time facilities, organized international exhibitions and later equipped the cities with electricity. [color plate 6]

With this industrial program the military principle of international competition established by the napoleonic era was to be surmounted by a civil competition of free market forces occurring within the framework of democratically organized societies.

Architects and Engineers of the First Industrial Revolution in France

France's industrial development after 1815 motivated the establishment of the École Centrale. Yet, in comparison to England, France's industry was behind by decades: consider for instance the "Iron Bridge" built over the River Severn near Coalbrookdale (1775–1779) and compare this with the Pont des Arts (1801–1803) or the Granary in Paris (1811), for example. The first industrial complex that provided the impetus for further developments entailed the greenhouses and glass pavilions built by the "Polytechnicien" and student of Durand, Charles Rohault de Fleury, in the Jardin des Plantes in Paris (1833).

In the second half of the 1850s France's industry had already reached the same level as that of Great Britain – compare for instance the "state of the art" represented by the palm house in Kew Gardens (Richard Turner, 1847) and the "Crystal Palace" (Joseph Paxton, 1851) with Victor Baltard's "Halles centrales" in Paris (1853, together with Félix Callet), or Eugène Flachat's parallel design of the halls (1850). It can even be maintained that the prestressed trusses already invented by Polonceau in 1837 and used in French railway stations represents a substantial step beyond the framework support structure built for the Crystal Palace.

"Halles centrales": The First Major Industrial Site in Paris

The "Halles centrales" by Victor Baltard and Félix Callet, which was begun two years after Paxton's "Crystal Palace" (1853), illustrates the high level of standardization, material technology and assembling technology that had been attained. Moreover, its spatial arrangement and the urban quality of the entire complex that exuded an architectonic presence not only visible on the outside but particularly also from within the interior, was created by the combination of iron support technology, natural light sources and ventilation, and represents a high point in the history of construction in France – the cultural value and lyrical quality of which was documented by a prominent eye-witness, Émile Zola, in his work "Le ventre de Paris".

Les Halles centrales in Paris by Victor Baltard and Félix Callet (Project: 1853; execution 1st stage: 1854–1857, 2nd stage: 1860–1866; torn down: 1971). (Marrey, 1989, p.41)

Émile Zola: "Le ventre de Paris" (1873). "And in their grand tours, when all three, Claude, Cadine and Marjolin roamed about the market halls, from every corner they perceived a portion of the cast-iron giant. It was these sudden glimpses, these unfamiliar architectures, the same field of vision that continuously appeared in different aspects. Especially in the Rue Montmartre Claude turned to look behind him when they passed the church. The tilted view of the far-off market halls thrilled him: a large archway, a high gate opened wide before him; then the two-storied halls towered above, their infinite Venetian blinds, their enormous awnings; impressions of profiles of houses and palaces standing above each other, a metal Babylon of Indian lightness pervaded with hanging terraces, hovering promenades and flying bridges thrown over nothingness. Again and again they returned, to this city, to roam about it without taking more than one hundred steps away from it. They returned to the warm market hall afternoons. Above the Venetian blinds are closed, the awnings rolled up. In the covered streets the air is listless and gray, intermittently cut by yellow rods produced by sun spots falling from the high windows. Muffled sounds come from the market. The footsteps of the few still at work echo on the pavements while porters with their metal tags sit in a long row on stone settings in the corners of the halls, take their shoes off and nurse their hurting feet. It is the peace of the sleeping giant, where from time to time a cock's crow is to be heard rising up from the poultry hall."[4]

[4] Trans. from: Zola, 1873.

The Decisive Thrust: Railway Engineering

Early industrial complexes such as Schneider et Cie., or Creusot (founded 1836), the Soc. Cail (1846), the Atelier Fives-Lille (1861) or the firm Moisant-Laurent-Savey (1866), to be described further below, played a central role in the industrialization of France.[5] Most of them had been traditional businesses, ateliers, smithies, etc. that in part had already been established the 18th century. Dunham notes that after 1815 many of the larger firms employed English skilled workers who had been trained in the most recent smelting techniques, and that English businesses were also established in France in order to take advantage of the still open market in its infancy.[6]

Bringing French and English industrial progress into line with each other primarily took place in the area of railway engineering; most of the above mentioned firms were leading examples of it. At almost the same time as George Stephenson tested his legendary steam engine "Rocket" on the railway line between Manchester and Liverpool (28 October 1829), the École Centrale offered its first courses (3 November 1829). The first French train was also powered in 1829, namely the Chemin de fer de Saint-Étienne.[7] Already in 1831 the French railway pioneer Auguste Perdonnet was engaged by the new industrial school to teach mechanical construction in railway engineering; the first independent and comprehensive course on the subject of railway technology to be introduced by Perdonnet 1832/33 was a world premiere event![8] France, and particularly the new school reacted quickly to this new industrial thrust. A map of the French railroad system of 1855 shows already an "all-encompassing" railway network.

[5] Lemoine, 1986, p.298–305.

[6] An example of this is Joseph Bessy who arranged for 42 English families to settle in the industrial region around Saint-Étienne and to take on a leading role in ironworks or to set up businesses; cf. Dunham, 1955, p.129.

[7] Guillet, 1929, p.320; in Perdonnet's indexes (European and American railways), however, the connection Saint-Étienne-Andrezieux figures first (1827) and Lyon-Saint-Étienne second (1832); cf. Perdonnet, Tome I, 1858, p.274f.

[8] The older École des Ponts et Chaussées, founded in 1747, followed suit in 1833/34 with the first courses held by the engineer C.J. Minard; cf. Picon, 1992, p.350.

Railway stations represented an important area of experimentation for architects and engineers. Important inventions with respect to large spans as well as double-curved supporting structures were made here. Moreover, railway stations posed multifarious challenges, as they were the starting-point for novel experiences in travel and speed; yet they also provided urban spaces for social communication. This double function – reception hall "in the front" and passing-

A map of the French railway network around 1855. (Picon, 1992, p.358)

through halls "in the back" – resulted in many cases in the kind of typical representative forms exhibited by polarized disciplines of construction of that time: "Architectes" – mostly from the Beaux-Arts school – built the front sites, "Ingénieurs centraux" constructed the industrial halls above the tracks.

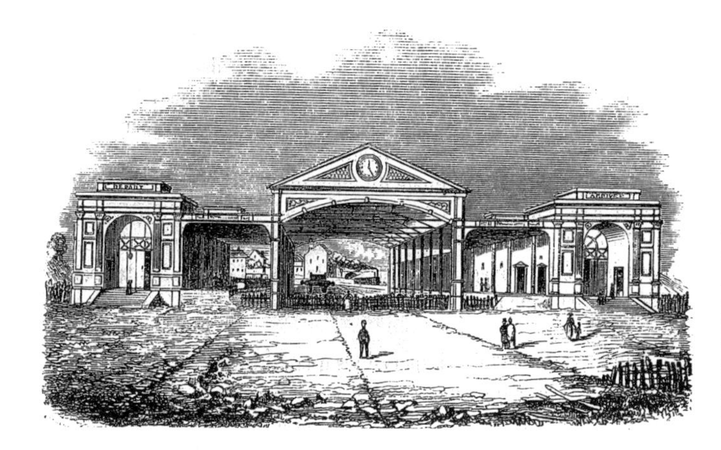

Gare de Versailles, the end-point of the Paris Line; Senior engineer: Auguste Perdonnet, first teacher of railway engineering at the École Centrale after 1830. (Perdonnet, 1860, p.186)

Heinrich Heine and the Age of Railroad Engineering. "What changes must now occur in our perception of things, in our imagination! Already the elementary concepts of time and space have been brought to waver. The railway has killed space and only time is left to us. In four and a-half hours you can travel to Orléans, and in just as many hours to Rouen. What will happen when the lines are built to Belgium and Germany and connected to the railway there! It is as if the mountains and woods have moved to Paris. I already smell the scent of German lime trees; already the waves of the North Sea break before my door."[9]

[9] Cited and Trans. from: Treue, Manegold, 1966, p.85f.

The fact that the newly founded École Centrale des Arts et Manufactures reacted to the industrial challenges of the time by offering comprehensive courses in railway engineering from the very beginning and by engaging one of the leading engineers of the day to teach at the school is not only an illustration of its leading role in the industrial process but also shows the interdependency between its novel industrial system of instruction and the history of technology. Many graduates of the École Centrale who received their diploma in this field, such as Polonceau and Contamin, had a great influence on the further development of French railway engineering.

The same can be said for the fields of industrial and mechanical engineering and of architectural engineering with its significant industrial structures, i.e. exhibition halls, etc. This will be treated further below.

Characteristic structures of transportation and goods supply, of production and hygiene, and also of leisure time and popular celebration defined the city skylines with their modern, industrial design. Lightness and transparency meant progress and modernity and exerted a certain amount of fascination of the kind Émile Zola described. At the same time writers such as Heinrich Heine were to be found who expressed their skepticism in the face of new developments.

Chronicle of the School's History

The first great era of the École Centrale treated here began with its inception in 1828/29 and ends with the school's transferal to government control in the year 1857. In these years the following founding individuals and initial faculty were active, such as Olivier (taught 1829–1853), Péclet (1829–1857), Ferry (1830–1864), Perdonnet (1832–1865), Mary (1833–1864) and Thomas (1835–1869) . The second generation of teachers and former “Centraux” included, for example, Polonceau (1851/52), Muller (1864–1889), Ser (1865–1888) and Contamin (1873–1893), who were active far into France's second industrial revolution. As will be elucidated at the end of this chapter, it was particularly in this phase that the school's teachers and students were to have a formative influence on the history of science, technology, industry and education – and not only in France.[10]

[10] The following chronicle is primarily based on the École Centrale's biographers: de Comberousse, 1879; Pothier, 1887; Guillet, 1929.

The Founding Phase: 1828–1831

This phase starts where a historical period ends; a period that saw Napoleon's fall (1815), re-established an "Ancien Régime" and suspended industrial development. The system of privilege that was reintroduced not only affected political but also economic civil rights and resulted in placing France on the fringe of developments, especially in comparison to Great Britain. The founders of the École Centrale also belonged to the circle of liberal forces and the industrial bourgeoisie who no longer wanted to wait for France to assume a position in the world appropriate to its culture. The motivating and formative individuals who founded the school included Théodore Olivier, Jean-Baptiste Dumas and Alphonse Lavallée.

Théodore Olivier (1793–1853). (Pothier, 1887, p.193)

This group attended the "Athenée", a club representing scientific and industrial culture, and read "Le Globe", a journal disseminating liberal ideas. Dumas lectured on chemistry at the "Athenée". The spirit cultivated here reflected the ideas of the Enlightenment and further provided them with a modernized, industrial dimension. The concept of industry entertained here stood for progress and was to serve not only the general welfare but also that of each individual. One of this movement's inspirational personalities was Saint-Simon. Olivier graduated from the École Polytechnique in 1811, was one of Monge's and Hachette's most favorite students, taught at the officer school in Metz (formerly Mézières) and continued Monge's Géométrie descriptive at the École Centrale. Dumas studied around 1816 at the University of Natural Sciences and Pharmacology in Geneva. In 1821 he moved to Paris and was appointed "Répétiteur" for chemistry at the École Polytechnique in 1824. Between 1849–1851 he held the office of minister for agriculture and trade and as such represented France at the World Fair in 1851 in London. Lavallée studied law and carried out this profession in Paris for a few years. He then worked as a merchant and trade advisor for his brother-in-law's plantation in the American South (Louisiana), returned to Nantes where he

Jean-Baptiste Dumas (1800–1884). (Pothier, 1887, p.313)

Alphonse Lavallée (1797–1875). (Pothier, 1887, p.263)

founded a school of agriculture and industry, and in 1827 he finally settled in Paris.

After the group's first meetings to establish the new school in October 1828, which they announced in "Le Globe"[11], ministerial authorization was given and the founding articles were formulated, which corresponded to a manifesto foresightedly drawn up by Lavallée already in 1827. In this manifesto Lavallée formulated the postulate that a basic course of studies must be completely pervaded by a synthesis of all sciences and that specialization should only occur in the second half of the course of studies. He stated further that the school should also contribute to scientific progress and that instruction should be made available to all interested and well-prepared young individuals free of cost and in a motivating manner.[12] Lavallée, as the school's first director and financier, signed a lease for an appropriate building complex after ministerial authorization was given: the Hôtel Jugnié (today's Picasso Museum).

[11] Journal "Le Globe", 1828, p.741 and 752.

[12] Pothier, 1887, p.22, as well as Doc.6, p.445.

Already one year later, on 3 November 1829, the École Centrale offered its first courses. The school's "Premier Prospectus", formulated during its preparatory phase, took up the Enlightenment idea of free and equal access to education for all citizens independent of their status, and declared its intention to continue the tried and proven concepts of teaching and methodology maintained by its sister school, the École Polytechnique, especially the "Méthode Monge" and the system of practice courses, although in an industrially expanded form. With this declaration the founding fathers of the École Centrale delineated the school's distinctive position in the spectrum of higher education for architects and engineers. The École Polytechnique's radical enrollment practice of that time (reduced number of candidates and limits to age of leave) allowed the school to open quickly and begin its activity with well-filled classes.

Its program of instruction was adapted to the needs of the time and took on the "English challenge". It was to train engineers and architects who, on a scientific basis, were to bring industrial production into motion, invent and perfect industrial processes in the various fields and manage modern businesses. More-

The Hôtel Jugnié, the École Centrale's first school building. (Guillet, 1929, p.16)

over, its "Esprit de L'enseignement" was pervaded by the conviction that all branches should form a "sole science" ("La science est une") as the basis for completing all technological, industrial and social assignments.

The school opened its doors and began its courses with 140 students. Entrance exams were only held in Paris, although with the cooperation of foreign experts. From the very beginning many foreign students were enrolled in the school. A large portion of the student body, 110 to be exact, were recipients of "half stipends" which were to help poor families and candidates from the least developed areas of France.

The July Revolution took place in the school's very first year in 1830. The school and its members considered it as a matter of course to voice their opinion in favor of ending the Restoration Regime. On the 27th of July 1830, the first day of the revolt, students of the École Centrale put on uniforms and assembled in the school yard and placed themselves at the command of a fellow student, M. Veret, who was also an officer of the Swiss army. They took on defensive positions in the neighboring city quarter and fought side-to-side with the Polytechniciens. The July revolt brought the Restoration to an end and established a modern government under the "citizen-king" Louis-Phillippe, Duc d'Orléans. The Centraux contributed as much to this transformation and thus learned an important civic lesson on freedom, equality and independence, which taught them to appreciate the value of their school's independence from government control and the value of their later professional status.

After the July events and following the new government policy encouraging industrialization, the École Centrale underwent a curricular restructuring in that a more in-depth industrial training was to become obligatory for the third school year. The school administration replaced the first model, "two plus one", meaning two years of basic and one year of specialized training, with "one plus two" stages: one basic year course (learning the general theory of sciences) was to be augmented by a two-year vocational course (practical training through exercises). This not only corresponded to the wish of industry but also that of the students. The in-depth training provided in the second and third year-courses reflected the Centraux' major fields of occupation: mechanical engineering (Machines et arts mécaniques), structural and architectural engineering (Constructions et arts physiques), inorganic chemistry, mineral resources and metallurgy as well as organic chemistry and agriculture (the latter was dropped as a specialized field already in 1832).

The curricular restructuring also entailed the introduction of a council of studies (Conseil des études) as well as a supervisory council (Conseil de perfectionnement) besides the actual school council or council of founders (Conseil supérieur or Conseil de l'École). While the latter was responsible for the school's general fate, for the choice of professors, Répétiteurs, personnel and school enrollment, the council of studies was responsible for issues of methodology and teaching as well as for matters of discipline; the supervisory council, based on the École Polytechnique example, ultimately supervised the implementation of the teaching programs, the maintenance of the educational goals as formulated, and carried out an analysis of the efficiency exhibited by the didactic model of instruction. The supervisory board was made up of important personages from the scientific, technological, political – and in contrast to the École Polytechnique – industrial communities. Regular "Rapports" on the achievements of the different departments were compiled and published.

In the second year of the school its founders decided, as was also done at the École Polytechnique, to introduce a "Chef d'étude" or a student representatives. They were appointed by the regular student body, whereby expertise and interpersonal characteristics such as goodwill and sociability played an important role. The representatives were responsible for discipline in the drawing studios and acted as tutors, which meant that they were to reinforce the learning material and in cases where the subject matter was not understood, explain the material or provide further help.

The Establishment Phase: 1831–1836

In the spring of 1832, at the time the École Centrale was about to refurbish and expand its courses, the city of Paris was hit by a cholera epidemic. The director and his family as well as some teachers and students were affected. Lavallée ordered the closing of the school and the students' return to their homes. The interrupted courses were to be taken up again in the summer holiday months.

The establishment phase is characterized by the introduction of novel industrial courses on railway engineering (Perdonnet, as of 1832/33), steam engines, turbines and hydraulics (Colladon, as of 1830/31) as well as industrial architecture and engineering (Mary, as of 1833/34). This will be treated further on. The initially popular English language course had to be abandoned in 1832 because of lack of demand; industrial hygiene suffered the same fate (Hygiène industrielle), which was re-introduced in the regular course plan in 1885; the most important course on industrial economy, which was planned from the very beginning, was further delayed and not included in the curriculum until 1856. These courses particularly expanded the students general educational horizon, the planning and introduction of which were the subject of continuous programmatic debate. The Swiss Polytechnical School in Zurich was then the first school to institutionalize these courses of general education as a separate department from its inception in 1855.

In this phase chemistry became increasingly important and was to further pervade the entire curricu-

A comparison of the École Centrale and École Polytechnique curriculum (1830/31). (Pothier, 1887, Annexe no.2, p.419)

ANNEXES. TABLEAU SYNOPTIQUE DES PROGRAMMES DE L'ENSEIGNEMENT N° 2

DE L'ÉCOLE POLYTECHNIQUE (1799) (*Extrait de l'Histoire de Fourcy*).

PREMIÈRE ANNÉE	NOMBRE des LEÇONS.
Analyse pure et appliquée.	120
Géométrie descriptive.	126
Éléments de machines.	27
(1) Physique et Histoire naturelle, une leçon chaque décade aux deux divisions réunies.	
Chimie théorique.	60
TOTAL.	333
Dessin topographique, levés sur le terrain.	29
Dessin de la figure et du paysage.	120

DEUXIÈME ANNÉE	NOMBRE DES LEÇONS.
Analyse.	48
Mécanique.	72
Travaux publics ou constructions.	54
Fortification.	54
Architecture.	45
Travaux de mines.	27
Chimie appliquée aux arts.	60
TOTAL.	360
Dessin de la figure et du paysage.	120
Chimie expérimentale (manipulations).	60

DE L'ÉCOLE CENTRALE DES ARTS ET MANUFACTURES (1830-1831) (*Extrait du Prospectus de 1830-1831*)

PREMIÈRE ANNÉE		NOMBRE des LEÇONS.
Analyse et Mécanique.	MM. DE CORIOLIS.	70
Géométrie descriptive.	OLIVIER.	70
Chimie générale.	DUMAS.	70
Physique.	PÉCLET.	70
	TOTAL.	280
Manipulations, une par semaine.		
Dessin, 3 séances par semaine.		

OBSERVATIONS

Ce Programme est extrait des prospectus publiés en 1830 et 1831. Il a reçu en 1832 et années suivantes les modifications signalées tableaux 4, 5, 6.

Le cours d'Économie industrielle n'a pas été professé. Le nombre des leçons fixé en principe n'a pas été toujours exactement appliqué. La durée de la leçon a été de 1 heure et demie et de 2 heures. M. Colladon donnait ses leçons de 2 heures au moins.

DEUXIÈME ANNÉE		NOMBRE des LEÇONS.
Théorie des Machines.	MM. WALTER DE St-ANGE.	30
Construction —	FERRY.	30
Constructions civiles.	GOURLIER.	36
Géométrie descriptive appliquée.	OLIVIER.	70
Physique industrielle.	PÉCLET.	70
Minéralogie et Géologie	Constant PRÉVOST.	36
Chimie industrielle et analytique.	DUMAS et BUSSY.	35
Histoire naturelle industrielle.	BRONGNIART.	18
Économie industrielle.	GUILLEMOT.	18
	TOTAL.	343
Manipulations.		

TROISIÈME ANNÉE		NOMBRE des LEÇONS.
Théorie des Machines.	MM. WALTER DE St-ANGE.	30
Construction —	FERRY	30
Travaux publics.	RAUCOURT.	36
Théorie des Machines à vapeur.	COLLADON.	22
Métallurgie { H^{ts}-Fourneaux	WALTER DE St-ANGE.	35
Métallurgie { Fabric. du fer.	FERRY.	35
Exploitation des mines.	PERDONNET.	36
Chimie industrielle.	DUMAS ET BUSSY.	35
Histoire naturelle industrielle.	BRONGNIART.	18
Hygiène industrielle.	PARENT-DUCHATELET	18
Statistique industrielle	GUILLEMOT.	18
	TOTAL.	313

(1) Les leçons de physique ne sont pas précisées; elles avaient lieu dans la matinée du 5e jour de chaque décade. Elles étaient communes aux deux divisions réunies.

lum where the Géométrie descriptive played an equivalent role at the École Polytechnique.

While general and analytic chemistry belonged to the basic course of studies, the "Chimie industrielle", representing a discipline of modern application in different industrial fields, was maintained as an independent upper-class course of study in which a diploma could be acquired. In addition, mechanics and metallurgy, mechanical engineering, structural and architectural engineering were also pervaded by a modern grasp of material chemistry. Thus, the use of iron structures and concrete in building required precise knowledge of material and processing, which was primarily acquired through the science of chemistry. The school historian de Comberousse described the methodological "penetration" of chemistry in the instruction programs of all disciplines in the École Centrale as a form of encyclopedic systematic ("Encyclopédie raisonnée").[13]

[13] De Comberousse, 1879, p.96.

The Consolidation Phase: 1836–1840

Thanks to the great number of graduates who spread the school's good reputation, the École Centrale became well-known throughout the world. Instruction was consolidated after the reform of 1832, the school organization found a definitive settlement in the institutionalization of a dual administrative and educational direction, and the school's finances improved. In the 1839/40 school year the school restructured the curriculum and formed two equally long courses of study: basic education was as a consequence extended to three semesters and the specialized course of study was reduced to three semesters.

In 1837 the school also altered its entrance policy. Besides verbal testing candidates were also required to write an essay on a geometrical and literary theme each. In addition, study papers, sketches and drawings also had to be presented that were completed at preparatory schools. In the opinion of the school direction, the highly qualified program of instruction offered at the school could only be effective if the student was appropriately prepared and intellectually capable, as entering the school represented the first step in a chosen career.[14]

[14] De Comberousse, 1879, p.89.

The government, in particular its respective ministry for public services, agriculture and trade, increased the stipend fund and considered its financial support not to be a policy of preferential treatment but one of motivation for those who could appreciate having a modern, highly qualified and specialized state-supported institution of education contribute to their career chances.[15]

[15] Prospectus of the École Centrale 1836–1838, in: Pothier, 1887, p.160.

Between 1836 and 1840 the school increased its progressive industrial status, an example of which was the course in railway engineering (Perdonnet), or the course in turbines and steam engine construction (Colladon), or Mary's structural and architectural engineering course.

The basic drawing course was also expanded. The École Centrale, as an industrial school, endeavored to probably compete with the Beaux-Arts school in its drawing classes by further developing the art of rendering with the help of basic systematic and technical principles. For this end it engaged teachers, like the École Polytechnique, who were graduates of that academy; the first was M. Thumeloup. This very basic decision, which incidentally is still practiced by most all of the polytechnical and industrial schools today, augmented the technique of drawing with a creative horizon and led to the evident improvement of rendering techniques in work submitted for diplomas stored in the École Centrale's archives.[16]

[16] Victor Contamin's diploma will be treated as an example further on.

Finally, the school reinforced its industrial direction by introducing vacation work (Traveaux de vacance), which became a part of the obligatory instruction. They entailed a comprehensive analysis of industrial or commercial businesses that students were to complete during their vacation, which included interviews, protocols, drawings and reports.[17]

[17] These will be treated (in the section on William Le Baron Jenney) more in detail and explained in the form of an example (cf. the introductory chapter).

Apex and Transferal to Government Control: 1841–1857
Curricular improvements of this period include "chemical technology" (as of 1844) and "industrial economy" (planned since 1829 but not realized until 1856); this period included Léonce Thomas' professorship (steam engines, 1835–1869) and well as Camille Polonceau's teaching activity in the field of mechanical engineering (1851/52).

The most important issue discussed from the middle of the 1840s was the École Centrale's relationship to government, whose role in encouraging industrial development and supporting the school was to be seen in a new light. As of 1846 the founders of the school worked towards transferring control to the state.

However, political conditions once again shifted during the February revolt of 1848 in Paris. The adverse effects of the rapidly progressing industrial expansion gave rise to socialist forces, who, because they were not able to voice their protest in parliament due a restriction in voting rights, manifested their protest on the streets. Between the 22nd and 24th of February 1848 the French capital again witnessed barricade warfare in which, as in 1830, Polytechniciens as well as Centraux fought on the side of the liberal republican forces. While the "Ingénieurs militaires" wore uniforms decorated in red, the "Ingénieurs centraux" differed from their colleagues at the École Polytechnique – not only by their voluntary activity – but by wearing uniforms decorated in blue. The February Revolution resulted in Louis-Philippe's abdication and in the declaration of the Second Republic; an interim government was established under the leadership of the author Lamartine and the scholars Dupont and Arago, the latter a Polytechnicien and Monge's successor. This government assembled, together with the socialist opposition, on the 25th of February.

A "right to work" formulated by this kind of political constellation for the first time was to be realized in practice by means of "Ateliers nationaux", which were to reduce unemployment. Immediately after the February revolt these government workshops mobilized approximately 120,000 unemployed workers who, under the guidance and leadership of many graduates of the École Centrale and the École Polytechnique, were to carry out work commissioned by the public sector (25 February to 26 May 1848). The person appointed by the government to oversee the "Ateliers nationaux", Émile Thomas, hoped to count on the good moral and disciplinary example set by the 75 Centraux, who were recruited by the government as leading engineers and technical cadre for this project.[18]

[18] However, it turned out that the government could not commission these workshops for truly necessary work, which as a consequence had a demoralizing effect and the experiment had to be broken off after just three months. After the initial euphoria and strong commitment on the part of most of the École Centrale school body, professors, students and graduates alike, the school was not able to return to its normal instruction activity until after the "Ateliers nationaux" were closed; Thomas, 1848.

The very first free and general election to be held on 23 April were to show, however, the radical and socialist forces' relative weakness; in the new National Assembly they were but a minority. Unlimited suffrage as well as the freedom of association, of assembly and of the press were established in the constitution passed on November 4th, however, a "right to work" was not. The fact that radical and republican forces were broadly rejected also nurtured the "Bonapartist cult", now centered around Louis Napoleon Bonaparte, the first Emperor's nephew, which swiftly began to take on a more clearer form. Thanks to his popularity, Bonaparte was elected into the newly established office of president on 10 December 1848 and, like his ancestor, immediately began to consolidate and expand his power. On 2 December 1851 he staged a coup d'etat and on 14 January 1852 his sole claim to power was confirmed by plebiscite; as Napoléon III he was crowned Emperor of the French on 2 December 1852.

In 1857, almost ten years after the February Revolution and under the conditions of the new "Empire" and the Paris of Haussmann, an era came to an end for the École Centrale: This private institution, which sought to carry out its work independent of political fluctuations, was now transferred to government. The preceding negotiations endeavored to guarantee the school's continuation according to the principles of its founders, to secure its model of instruction, its educational purpose and to receive acknowledgment for the specialized work that had characterized this modern school.

The contract regulating the government's responsibility for the École Centrale (13 April 1857) fulfilled the three of the school's major conditions: to pay a takeover fee as compensation to Lavallée, which represented his pension, to pay pensions for all other professors and finally to guarantee that the school's entire expertise, accomplishments and means ("Produits de L'École") would not be appropriated but used exclusively for the school's benefit and for paying its debts.[19] In other words, the principles that contributed to the school's success were to be maintained.[20]

[19] Pothier, 1887, p.232–247.

[20] De Comberousse, 1879, Preface p.XII.

In looking back from an historical perspective it becomes evident that the school lost its impetus and initiative with its transferal to government control and that it was replaced in its leading industrial role by other modern schools in Europe and America. At any rate, it was in the era between 1829 and 1857 that the École Centrale particularly contributed to the establishment of new schools such as the Rensselaer Polytechnic Institute, the Polytechnical School in Karlsruhe and – indirectly – the Swiss Polytechnical School in Zurich by the development of its pioneering and exemplary model of instruction.

A further fruit of the École Centrale's flourishing era is represented by the founding of the "Société des Ingénieurs Civils" in 1848 as well as the journal "Le Génie Civil".[21] These instruments of organization not only served to encourage the concept of solidarity, to provide information and education, but it also served to formulate the professional concept of the modern architect and engineer by publicly assuming responsibility and by advertising their ingenious engineering activity as contributions towards social and cultural progress, thereby disseminating the new spirit.

[21] For further elucidation cf. the section on placing the school in the context of the era further below.

"Industrial Sciences" – A Concept of Teaching and Methodology

The decisive innovation, yes even turning-point resulting from the establishment of the École Centrale des Arts et Manufactures in the year 1829 is to be found in its conception of the "industrial sciences" (Sciences industrielles) as a methodological instrument in establishing the correlation between basic theoretical subjects and practical exercises of application. This school, for the first time in the history of the education of architects and engineers, adopted a novel model of instruction as a reaction to the industrial challenge that had increased in intensity since 1815, thus contributing to raising the standards of manufacturing and industry. From this a dynamic relationship between education and technology was established.[22]

[22] This will be further explained by the example of individual student biographies.

What Does the "Industrial Model of Instruction" Entail?
It was primarily concerned with overcoming the

Géométrie descriptive as the sole methodological thread between all fields of engineering by adding intermediate sciences, the "Sciences industrielles", i.e. industrial chemistry, industrial physics, mechanical engineering, structural, architectural and industrial engineering, etc. By introducing a canon of application-oriented subjects that lay between pure basic theory and practical exercises or application courses (projects), it was possible to include in the course of studies from the very beginning information on the current state of technological development and to generate an interest in modern praxis, or, in the tasks demanded by the time. Once again, the ideas of the Enlightenment were formative here in the sense that scientific and technological developments and achievements were to contribute to the general well-being. Issues such as hygiene facilities in hospitals and urban housing complexes were introduced in architecture classes, for instance by Mary's successor Émile Muller, as project assignments on living comfort.

Industrial housing ("Immeubles industriels") as a project theme in Émile Muller's architecture class (1875); sanitary and hygiene facilities in apartments (Student: Delmas). (ECAM; Dossiers F.Hamon, Paris©)

As a result, socially useful work and the goal to create social goods were integrated into the course subject-matter and thus became established as principle aspects of the instruction model. This had already found expression in the declarations formulated by the school's founders, who sought to carry over the basic ideas of the Enlightenment into the industrial era through the school's graduates, who, educated in this manner, would carry this spirit out into the world in their professional work. As François Pothier, the school's chronicler, emphasizes, it is not enough to just invest capital in the realization of major works to satisfy social needs, it is equally important to count on the support of the sciences, and industry can only boldly progress if it is "enlightened and knowledgeable".[23]

[23] Pothier, 1887, p.36.

A committed industrial approach did not mean, however, that the Géometrie descriptive lost its importance: it was still conspicuously taught by Théodore Olivier, school founder and Monge student. Yet it was no longer a key science, as the industrial sciences took first place. In addition, mathematics further developed geometrical concepts as instruments for calculating support structures, and as a form of "graphic statics" they paved the way for iron and metal construction.

A second characteristic of the industrial model ("Modèle industriel") as a modern principle of education for architects and engineers is that it represents a further development in learning theory, whereby the Sciences industriels alleviated the transfer between theory and practice and illustrative material was used derived from current technological and industrial practice: reality, therefore, no longer was presented as a model orchestrated in the school classroom nor as a finished product in an urban situation, as was the case in Durand's polytechnical architecture course. Explaining theory by means of true examples in Charles-Louis Mary's architecture course differed from that of the École Polytechnique in that construction sites were visited and unfinished products were examined; the students' activities, therefore, addressed working processes, production and assembly procedures at the level of actual construction activity. In contrast, the Polytechnical School sim-

Classes at the École Centrale; drawing studio and auditorium (second generation Centraux around 1880). (École Centrale, 1981, p. 72)

ulated and condensed reality in the auditorium and laboratory, as a reflection of reality; the kind of experimentation material it used was partly derided by the Centraux as "playthings" ("Joujous").

School and Industrial Practice
The combination of all of these forms of teaching and learning produced an attractive, concentrated and continuous course of study. Scientific principles and industrial sciences were linked together, similar to an encyclopedic "Tableau"; in this way a standardized study for all students was put together. While the first three semesters addressed all subjects, first basic courses (two semesters) and then an introduction into applied science in the third, in the second half of their

studies the Centraux concentrated on an additional special discipline of their choice. In this way a full course of studies was completed. A graduate of chemistry could also easily work in metallurgy (architecture and construction/architectural engineering), as did Gustave Eiffel, for instance (this will be treated later on). As a consequence, the school's diploma reflected an exceptionally high demand placed on general as well as specialized knowledge.

Thanks to this novel form of instruction students quickly came into contact with the current state of modern industrial practice, which was of course particularly fascinating to them. Since some of the most important pioneers of the most progressive fields of science and technology of that time taught at the school as well, the students were given the feeling that they were participating in the pioneering achievements of industrialization themselves, and were convinced of becoming research scientists, inventors or leaders of industry after their studies and of carrying the modern spirit out into the world. Not only was spreading the school's reputation part of the "industrial corps spirit" but also pride in the achievements of the French nation.

Modern Industrial Courses of Study: The "Modèle Industriel"

Examples of visits to construction and production sites as well as of laboratory experiments, which were used as modern forms of teaching and learning in the course of instruction, will be treated in the following according to the applied courses offered by the school's leading Professors Perdonnet, Colladon and Mary. They were of formative importance in the development and success of the Modèle industriel.

Perdonnet's railway engineering course. Auguste Perdonnet (1801–1867), whose father came from the Swiss Canton Waadt, studied at the École Polytechnique. However, in 1822 he was dismissed from the school for political activity (together with Léonce Reynaud and their entire drawing class); the Restoration Regime did not tolerate opposition at the "École" (this pertained to the "Carbonari" movement imported from Italy which impressed broad sections of the middle class, found approval among intellectuals and gained ground in the army; "Polytechniciens" were especially active and enthusiastic followers of the "Charbonnerie"). Perdonnet subsequently traveled to England in order to continue his studies in metallurgy. Even before the July Revolution of 1830 he wrote the first French treatise on railways ("Mémoire sur les Chemins à Ornières", 1829). After his return in 1830 he and his fellow students were rehabilitated by Louis-Phillipe's new government.[24] Already in 1831 the École Centrale appointed him as professor and established an independent and complete course of railway engineering studies as of the school year 1832/33 – the first of its kind world-wide! In 1838 he took overall control of the Chemin de fer de Versailles and in 1845 became director of the Chemin de fer de l'Est. As of 1862 he was also director of the school. Perdonnet wrote many papers, including a report on English railway engineering (1827) and most importantly a comprehensive textbook in two volumes (1858 and 1860).[25] As school director he compiled a report on the École Centrale in which he compared it to the Swiss Polytechnical School in Zurich.[26]

[24] Pinet, 1887, p.115–117.

[25] Perdonnet, 1858 and 1860.

[26] Bissegger, 1989, p.129f.

Auguste Perdonnet, Portrait. (Guillet, 1929, p. 38)

Perdonnet put together and standardized the railway engineering course from scratch. In his course he compiled and included within just a few years a spectrum of organizational, economic and legal issues that had yet to be solved and that were mostly derived from his practical experience. Once again theory and practice were linked, theoretical knowledge and practical application were manifested in Perdonnet's course as a form of industrial science. As Perdonnet remained completely informed about the progress and problems of his railway line, he took the advantage of planning excursions with his students in order to establish the methodical correlation to his instruction. As F. Pothier, chronicler of the school and student of Perdonnet, mentioned, he also at times made his students march along the entire line between Paris and Sèvres in order to view the workshops and depots including the wagon pool. Pothier emphasized how important these excursions were as illustrations of the material treated in Perdonnet's course of instruction.[27] Other excursions led to the locomotive production sites at Chaillot and to new bridge works such as the hanging bridge on the Île St-Louis built by Seguin.[28]

[27] Pothier, 1887, p.197.

[28] Weiss, 1982, p.132.

Construction site of the railway line Paris-Versailles near Clamart, popular excursion site for Perdonnet's railway course. (Perdonnet, 1858, Pl.6, p.371)

Combining lectures oriented on theory with a practical framework of reference and visits to actual construction sites is a form of teaching and methodology that substantially augments the polytechnical principle of instruction by a further industrial dimension. Perdonnet's type of instruction swiftly and decisively established the Modèle industriel's competency as an educational form and produced the "Ingénieur industriel". In the eight classes held by Perdonnet many successful railway engineers of the pioneer era were trained who continued to evolve and invent further technological and organizational aspects and who established or managed firms throughout the world. On the average, during Perdonnet's years as a teacher, one in five students received their diploma in railway engineering.[29] As of the 1840s the possibly most important institution for educating railway engineers in France besides the École Centrale was also the École des Ponts et Chaussées.[30]

The course on mechanical engineering held by Jean-Daniel Colladon. Colladon was born in 1802 into an old and wealthy family in Geneva.[31] Already in Voltaire's time and also at the beginning of the 19th century, thanks to its university and the Académie des Sciences, Geneva was an important city for scientific circles. Let us call to mind Dufour's activity in this city. Colladon visited the Collège and was interested in the natural sciences and technology, especially in physics

[29] Picon, 1992, p.601.

[30] Cf. in detail: Picon, 1992, Chap.7 (Section: "Les difficiles débuts des chemins de fer français").

[31] For biographical information, cf. Mallet, 1893.

and chemistry. One of his friends was Jean-Baptiste Dumas, later co-founder of the École Centrale, who studied at the University of Geneva. Around 1817 Colladon and a few of his friends founded a Société de philosophie; Dumas was also invited to their meetings.

In 1822 Dumas moved to Paris, an aspiration held by many young Genevans who were interested in science and philosophy. Colladon, after receiving his diploma in law in 1824, worked as an inventor and participated among other things in a competition held by the Parisian Academy in the year 1827. Supported by Arago he submitted a treatise on the compressibility of fluids, for which he won the Grand Prix. However, it was not until 1876 that the academy accepted him, first as "Correspondant" and finally in 1893 (as a 91–year-old!) as full member; because he was a Swiss national and therefore foreigner, he was placed on the waiting list for the limited seats ceded to foreigners. As a member he became M.N. Rigenbach's successor, who invented the rack-railway and built the Rigi Train near Lucerne.

In 1829, the year the École Centrale was founded, Colladon was already internationally acclaimed as an eminent scholar. His works include, for example, the construction of effective paddle wheels for the paddle-steamer "Wilhelm Tell", which sailed Lake Geneva. In 1828 Dumas invited his friend Colladon to take part in the preparations for founding the new school in Paris. Immediately after its inauguration Colladon took over the physics course and taught a novel course on steam engine construction.

He remained at the school only until 1836 and subsequently expanded his teaching position in physics and mechanics at the University of Geneva. Colladon was succeeded by his student Léonce Thomas (diploma in 1833). In the 1870s Colladon was commissioned by Louis Favre, senior engineer of the Gotthard Tunnel construction, to develop a drill cylinder powered by pressurized air and gas to cut through the Gotthard, which had important procedural advantages and increased the drilling speed.[32]

[32] Colladon, 1875.

At the École Centrale Colladon specialized in the construction of water turbines in mechanical engineering. Before each year-course began he acquired the most modern turbines and machines from Paris factories and transported them to the school in order to demonstrate their function to his students. He ordered, for example, 15 different pumps to be assembled in the laboratory, which he powered by water before the very eyes of his astonished students. This demonstration was followed by an analysis of the different types and qualities exhibited; the students then disassembled the pumps, completed detailed construction sketches and technical descriptions and then reassembled them.[33]

[33] Weiss, 1982, p.128f.

Such demonstrations not only had educational value in that unique experiences were imparted by a very committed and exemplary teaching personality, but they also corresponded to the novel understanding of practice that this modern industrial school entertained; they served to illustrate generally applicable laws of science that were treated in theory classes or lectures, they represented practical exercises and formed the basis for project work.

What was also new was the employment of original-sized machines in the school laboratory. In addition, Colladon organized excursions and visited workshops and factories; during these visits students were encouraged to take a "hands-on" approach, which according to Colladon should be a part of every young engineering student's experience at an industrial school. Colladon took the opportunity during these visits to train his students in observing the machines and mechanics very closely and in sketching them as precisely as possible from memory when returning to the drawing studio.[34]

[34] Pothier, 1887, p.473.

Teaching Architectural Engineering at the École Centrale and the Industrial Orientation provided by Charles-Louis Mary

On the History of Instruction in Architecture

In the first school year after the École Centrale was founded Gourlier put together a course in construction called civil construction or architecture ("Constructions civiles") comprising 36 lessons for the second year-course (1830/31). Already in the following school year Gourlier left the school and was replaced by Raucourt, who received his diploma under Durand at the

École Polytechnique in the same year as Dufour, Pichard, Hegner, as well as Mary. Raucourt belonged to the Corps des Ponts et Chaussées. The construction course was now divided into Constructions civiles (architecture) and public works or "Travaux publics" (roads, bridges, port facilities, etc.). After two year-courses, that is as of 1833/34, this educational subject was reorganized by the Polytechnicien Charles-Louis Mary, who completed his studies as engineer at the École des Ponts et Chaussées. His successor was Émile Muller (diploma in 1844 at the École Centrale), who took over the architecture course as of 1864. In 1872 a special course came into existence called "Éléments d'architecture", which was taught by Demimuid, graduate of the École Centrale (1858) as well as the École des Beaux-Arts; Muller meanwhile continued to teach Constructions civiles until 1889. His successor was Denfer. In addition, during Muller's years as teacher the subject of Travaux publics was taught as a special course by engineers of the Ponts et Chaussées (as of 1865).[35] Thus, in the last third of the 19th century the school instituted the division between a Beaux-Arts-inspired architectural teaching and civil engineering, thus establishing them as two different disciplines.

[35] Guillet, 1929, p.124f. and p.140.

Mary's Architecture Course

Mary (1791–1870) was an active and responsible supporting figure of the École Centrale. As one of Durand's students he was also a Polytechnicien of the first hour (diploma in 1808). In his function as administrative architect of Paris he organized the pioneering feat of the city's water supply. In 1833 he began his work as architecture professor at the new École Centrale. He also concurrently managed a well-known private office in Paris. In addition, Mary lectured on navigation systems, inland navigation, port facilities and other subjects at the École des Ponts et Chaussées(as of 1842).[36]

[36] Biographical recognition in: Pothier, 1887, p.291f.

After about 20 years of teaching Mary compiled a kind of encyclopedic compendium of the then current state of development of architectural engineering, which included an historical review and which as of 1852/53 he used as a textbook in his year-courses. The work was comprised of a handwritten lecture manuscript and a comprehensive collection of illustrative plates. Since different scripts exist it is to be assumed that Mary revised or up-dated his handwritten texts.[37]

[37] Mary, 1852/53.

Essential Features of Architectural Theory

Mary's "Cours d'Architecture" conveyed the École Centrale's basic principles of architectural theory that for the next thirty years of graduates were to determine the education of architects and engineers of the industrial era. In considering the students that studied under him, such as Camille Polonceau, Émile Trélat, Gustave Eiffel, William Le Baron Jenney, Armand Moisant, Victor Contamin and others, one can say that his influence was immense. With respect to the theory taught by his teacher Durand at the École Polytechnique he intensified the *structural aspect* of construction: "Architecture has composition and structure of buildings as its object." So reads the first sentence on the first page of his textbook. The equal emphasis of structure and composition paved the way for structural design's predominance over composition, which often, as Mary remarks, was equated with the decorative aspect, yes even with "architecture" as a whole. Mary understood structure to be the "art of execution", of erecting a building, determining the site and volume, deciding the proportions and structural features according to the material available or according to that which would best suit the building in nature and character.

With this approach to structure as the all-present factor in architecture and generator of design, the basic principles of "functional building" were anticipated a half a century before "Modernism" came into being. In this respect Mary belongs to the "forerunners of early modernism". One of Mary's students, William Le Baron Jenney, was one of the first to architectonically realize these principles of design – around 1880 in Chicago with the new generation of office buildings, or "Skyscrapers", after the "Great Fire" of 1871.

In the field of composition Mary also goes beyond Durand. It is not enough to take only the topographic layout and geometry of the site into consideration, but also climatic conditions and the different cultural traditions and needs; even building traditions within one country often show great variations throughout history. In this respect Mary in his teaching distanced himself from the kind of architectural teaching that encouraged the copying of architectural styles and reference buildings as part of instruction.

According to Mary, structural problems often follow eternally valid and exemplary rules, which the Roman arch is an example of. Similar to Durand, in his textbook Mary developed a "Marche à suivre", meaning a methodical series of steps in the process of design. However, the method is not limited to analyzing the products of old and new master builders to serve as models, and therefore also goes beyond Durand's establishment of "Tracés régulateurs", which places greater importance on the arrangement of basic spatial features according to a grid system. Although this forms Mary's point of departure, he went further to address structural typology and material technology.

Mary's Textbook "Cours d'Architecture"

Grid measurements were already determined according to Durand by spanning distances, meaning by structure and material, and were no longer purely proportional; what was therefore decisive in design were no longer the measurements of columns and the distances between them, as according to Vitruvius, but the structural (virtual) unit spacing.[38] Mary furthermore increasingly treated structural types and material characteristics in his architecture course. In 134 pages of his textbook titled "Knowledge of Materials" he explains the characteristics and the application of all available and tested materials such as wood, stone, brick, many types of mortar and lime as well as concrete. This is followed by a chapter covering 68 pages on structural systems, e.g. a typology of floor structures that includes the model of a double-curved hollow filler block floor held together by iron tie rods fastened onto the side walls. A high point of architectural teaching at that time is represented by the reference to and illustration of an iron girder type structure with double-T profiles. This structure was common in France after Mary (as of 1851) but was popularly used already around 1790 for textile factories ("cotton mills" of the Midlands in Britain.

[38] Cf. the chapter "Ende des Vitruvianismus. Durand" by Germann, 1980, p.243.

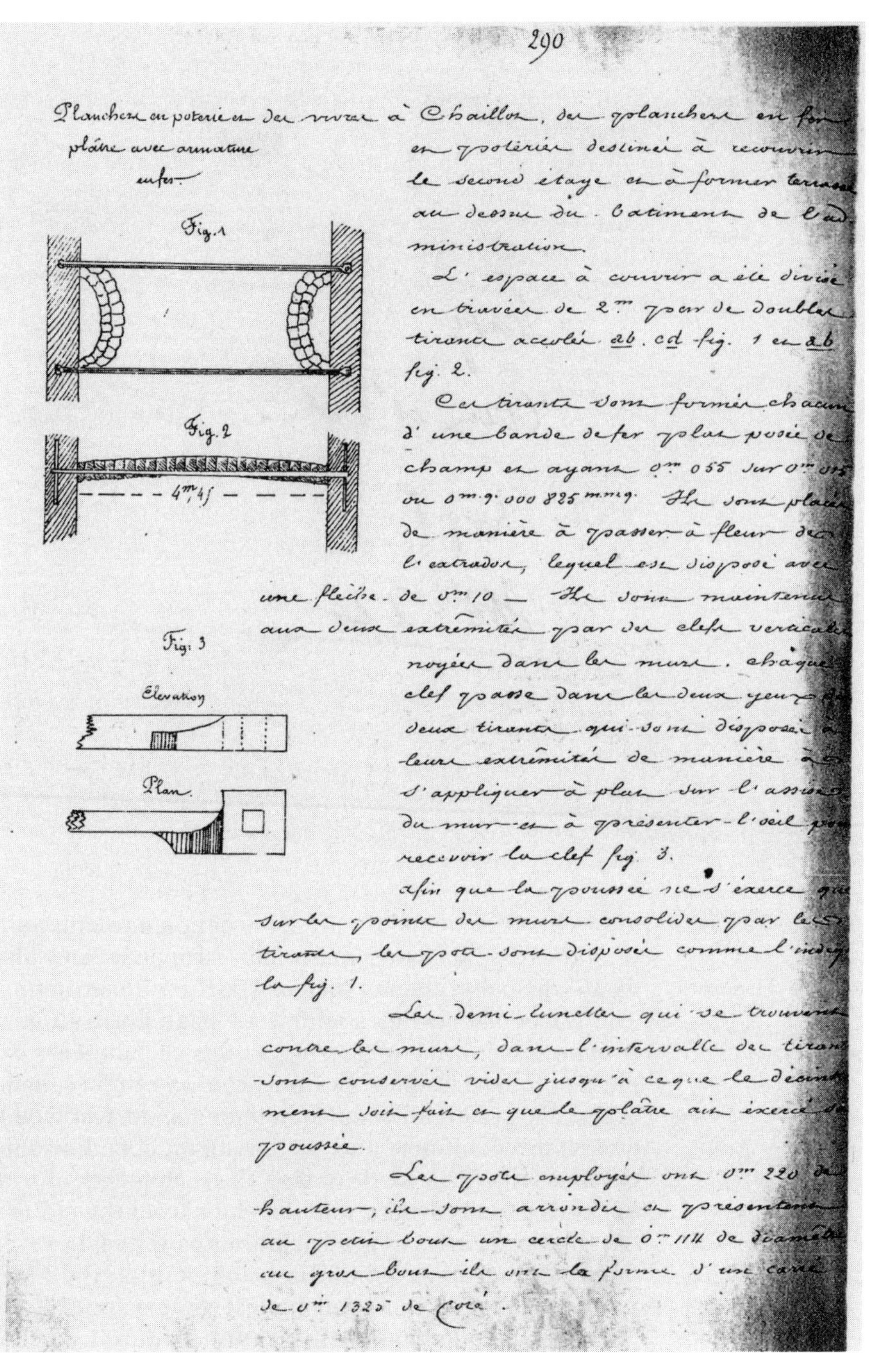
290

Plancher en poterie et plâtre avec armature en fer.

Les voûtes à Chaillot, des planchers en fer en poteries destinés à recouvrir le second étage et à former terrasse au dessus du bâtiment de l'administration.

L'espace à couvrir a été divisé en travées de 2m par de doubles tirants accolés ab. cd fig. 1 et ab fig. 2.

Ces tirants sont formés chacun d'une bande de fer plat posée de champ et ayant 0m 055 sur 0m 015 ou 0m.q. 000 825 mm.q. Ils sont placés de manière à passer à fleur de l'extrados, lequel est disposé avec une flèche de 0m 10 — Ils sont maintenus aux deux extrémités par des clefs verticales noyées dans les murs. chaque clef passe dans les deux yeux des deux tirants qui sont disposés à leurs extrémités de manière à s'appliquer à plat sur l'assise du mur et à présenter l'oeil pour recevoir la clef fig. 3.

Afin que la poussée ne s'exerce que sur les points des murs consolidés par les tirants, les pots sont disposés comme l'indique la fig. 1.

Les demi-lunettes qui se trouvent contre les murs, dans l'intervalle des tirants sont conservées vides jusqu'à ce que le décintrement soit fait et que le plâtre ait exercé sa poussée.

Les pots employés ont 0m 220 de hauteur, ils sont arrondis et présentent au petit bout un cercle de 0m 114 de diamètre au gros bout ils ont la forme d'un carré de 0m 1325 de coté.

Manuscript page from Mary's textbook with a sketch of an iron girder floor structure. (Mary, 1852/53, p.290)

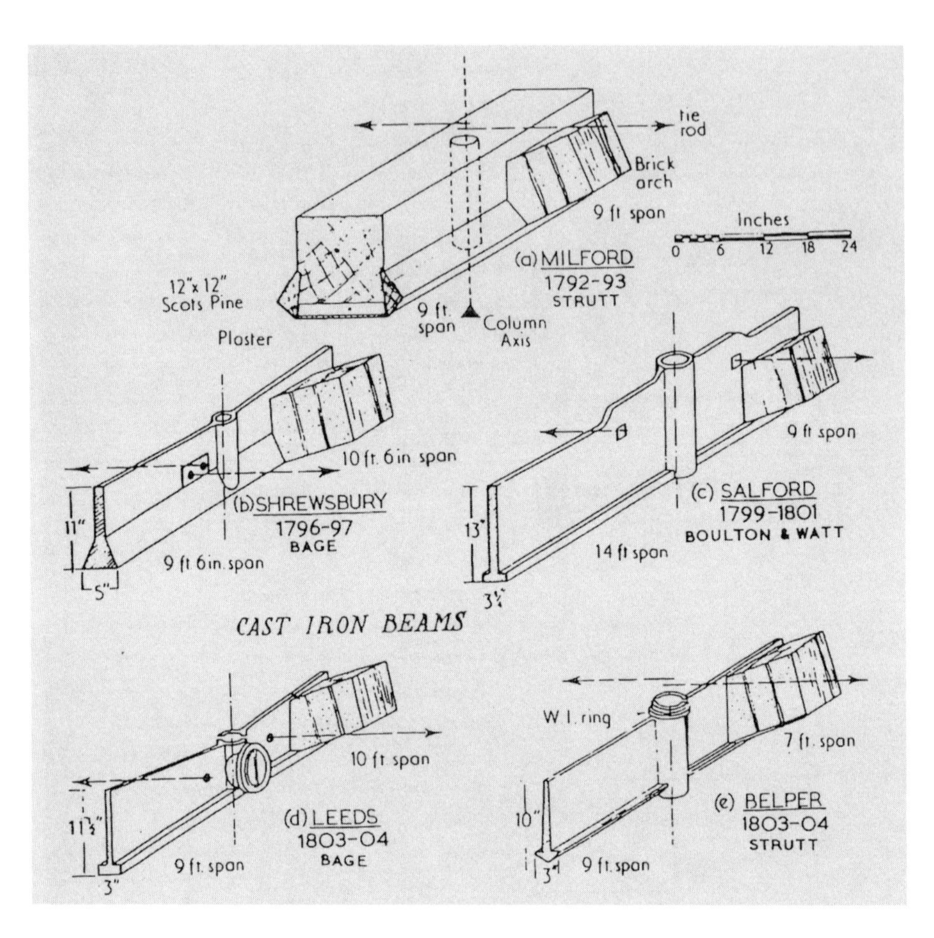

Typology of iron girder roof/floor structures used in the cotton mills of England's Midlands (1792–1804). (After Singer, 1975, p.477)

In his textbook Mary presents detailed calculations of statics for arcuated and other structures and also treats special problems such as staircase construction. The textbook concludes with a 43-page illustration of iron architectural engineering. In this section Mary explains the transition from iron's role as reinforcement to its role as a stabilizing element in masonry or wooden systems of binders and joists to an independent supporting structure; he describes the technology of iron anchorage and jointing and develops from the materials' characteristics new possibilities of application.

In the second part of his teaching material Mary published large formatted plates comprising a comprehensive collection of illustration material, including more recent iron structures such as the Granary's roof structure or other covered markets. In this con-

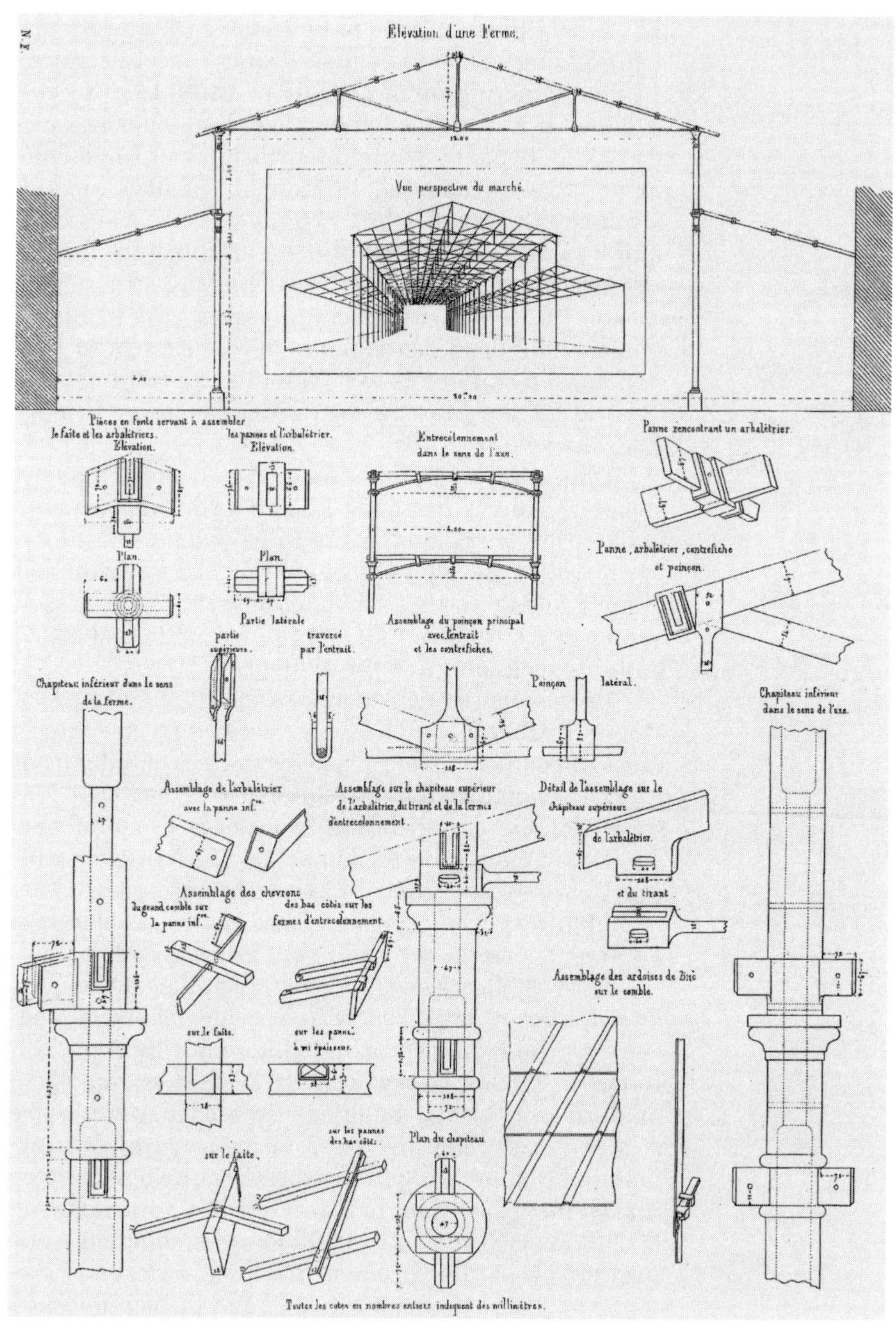

Structural table of the "Marché de la Magdeleine" in Paris (1824) from Mary's textbook (Plates). (Mary, 1852/53, PL.XLI.)

text the Halle aux blés is to be considered a prototypical building in the history of France's iron architectural engineering, built already in 1809–1813 by the architect François-Joseph Bélanger (engineering metalwork: F. Brunet). While the principles of wood construction were adapted to iron in constructing the Granary (afterwards the Exchange), later a decisive step was taken in structural iron architecture for designing covered markets: the supporting structure is broken up into cast-iron compression and wrought-iron tie members. An example of this change in iron structural technique is to be found in Mary's textbook illustrating the Marché de la Madeleine by Veugny from the year 1824.[39]

[39] Mary, 1852/53, p.344–348.

It must be noted that Mary treated these modern structural forms from the beginning of his teaching career. One of his students, Camille Polonceau (diploma in 1837), developed from within the environment of the École Centrale's school of thought a prestressed supporting structure that went down in the history of building technology as the "Polonceau Truss".

The treatise on construction takes up two-thirds of Mary's textbook, which clearly defines the teaching's focus: Structural theory provides the key to industrially oriented instruction in architecture at the École Centrale. Not only are structural types delineated but also detailed explanations of connecting and fixing technology as well as methods of procedure are given. *This teaching represents the link between or the common point of design for the architect and engineer.*

In 1873 Mary's second teaching material was produced, this time on road and bridge construction. Here we find the first iron bridge built by Abraham Darby (1775–1779) as well as further examples of modern iron bridge building, such as the Ponts des Arts, Pont d'Austerlitz, Pont de Fribourg in Switzerland (a hanging bridge), the Britannia Bridge by Robert Stephenson, and many more. In this textbook Mary not only presents the objects as complete entities but breaks them down into their structural system and components from a material-technological viewpoint and in great detail; the plates contain information on methods of procedure and assembly techniques.

Aspects of Teaching and Methodology
Mary's design theory entails a systematic step-by-step process and a sequence of program and building-site analyses, the establishment of spatial needs (for machines, workers, residents, etc.) and a novel view of the procedural unfolding of solutions by examining the variables in a process of optimization that endeavors to work out the building program in an interplay between the whole and its individual parts. With this method of design, completing the building assignment is not accomplished in a direct way; instead, possible solutions are analyzed and developed further in context. In this way Mary responds to the more complex problems posed by the industrial era and to the conditions of production that govern new materials such as iron and glass, which because of the characteristics they exhibit in support structures, connecting joints and in their load-bearing ability, are of major consequence to the initial functional and spatial conditions of the design.

In Mary's Marche á suivre a combinatorial analyses as well as the "Papier quadrillé" developed by Durand are used as tools of procedure. Mary recommends quick and initially less precise work in the design phase, when it is just a matter of developing useful solutions that are ready to undergo continuous scrutiny. Once a satisfactory solution has been found, the working method must proceed in an exact, definite and detailed manner with the help of the Papier quadrillé: "Precise studies are only to be completed after working quickly and approximately until a point has been reached in which by a satisfactory combination of all factors a solution has become possible."[40]

[40] Trans. from Mary, 1852/53, p.60f.

This method of working is conducive to correctly solving all kinds of building assignments in a systematic way and helps in making architectural design more operable: which means it can not only be learned by all but also can be universally applied. As an open system it is capable of being transferred to unfamiliar problems and as such is a vivid example of a Science appliquée.

Mary's designing method is on the one hand a further development of the "Méthode Durand" (specifically with respect to its step-by-step process of design), and it is on the other hand a precursor to the modern teaching and methodology of architectural instruction, the kind of which was introduced in the Architecture Department of the Swiss Federal Institute of Technoloy in Zurich since 1960.[41] Although little known, with his method of design Mary established an important paradigm shift in the process of design: from solution-based to *procedure-oriented design*, or, from working with building types to working with *problem patterns*.

[41] This "basic design course" will be mentioned again in the outlook at the end of this book.

Mary's design process is different from that of his teacher Durand in that it is not inductive, in increasing order from the individual to the whole, from the element to the system, but assumes knowledge of specific aspects such as territory, the building site's geometric form, climate, culture, spatial and room requirements for specific uses, also rules of combinatorial analyses, etc., and goes on to determine the entire surface available for the building complex by studying the variables involved and undertaking a combinatorial analysis of all individual parts until the most detailed studies have been accomplished, thus entailing a systematic and complex solution-finding process. Those designing were therefore methodically capable of grasping and working out the complexity of all aspects of design and structure "simultaneously". *Thus, "simultaneousness and complexity" in Mary's design and structural procedure further augment his teacher Durand's Cartesian inductive-deductive method of teaching and learning.*

Field and Vacation Work ("Travaux de vacances")

The founders of the École Centrale introduced a further didactic innovation in the form of field and vacation work, which as a an integral part of the school instruction was also obligatory.

Field and vacation work, in comparison to today's schools of architecture, made great demands on the students of the École Centrale at that time. Jenney's biographer, Th. Turak cites one such assignment from the year 1865: It concerned an analysis of an industrial factory according to organizational and functional criteria and to spatial and structural aspects, the machinery and their modes of operation; the correlation was to be made between energy input, production results, machine and apparatus volume and the structures enclosing them, the relationship between operative processes (quantitative and qualitative), operation and maintenance of machines, work organization (number of people and structure) and, calculated from this, the financial cost of production in relationship to the net price of the finished product, the number of workers and their salary in relationship to one thousand kilograms of the finished product. Moreover, during vacation work the factory's vicinity to the next larger city, its supply and development, and finally the conditions of health and security of the factory's workers and their working habits were also to be determined. The entire facility was to be observed, analyzed, recorded and sketched; interviews and explanations augmented the process. In conclusion, the students were to revise, demonstrate and present the documents collected as well as their basic knowledge.[42]

[42] Turak, 1986, p.33.

To illustrate the precise working method customary for the Travaux de vacances the following plate will include a drawing of the iron mill Creusot executed by the student F. Campionnet from the year-course 1864/65.

Placing Mary's Architecture Course in the Context of the Era

Modern iron architectural engineering, which Mary integrated into his classes and teaching material following the standard structures used for early English textile factories (since around 1790) and in consideration for the quickly expanding French industrial revolution, is one of the most innovative achievements of architecture of that time. Particularly this and the nascent concrete structural technique was to be conveyed to students of the "industrial polytechnical school", as the École Centrale was also called. Semester and final assignments mostly entailed work on industrial complexes, halls with the greatest spacing of supports possible, etc. (an example of this will follow). What was significant here was that students of all courses worked on issues of civil engineering and building constructions, since industrial processes (iron smelting, steel processing, chemistry, gas production, etc.) were always to be performed in functional building facilities.

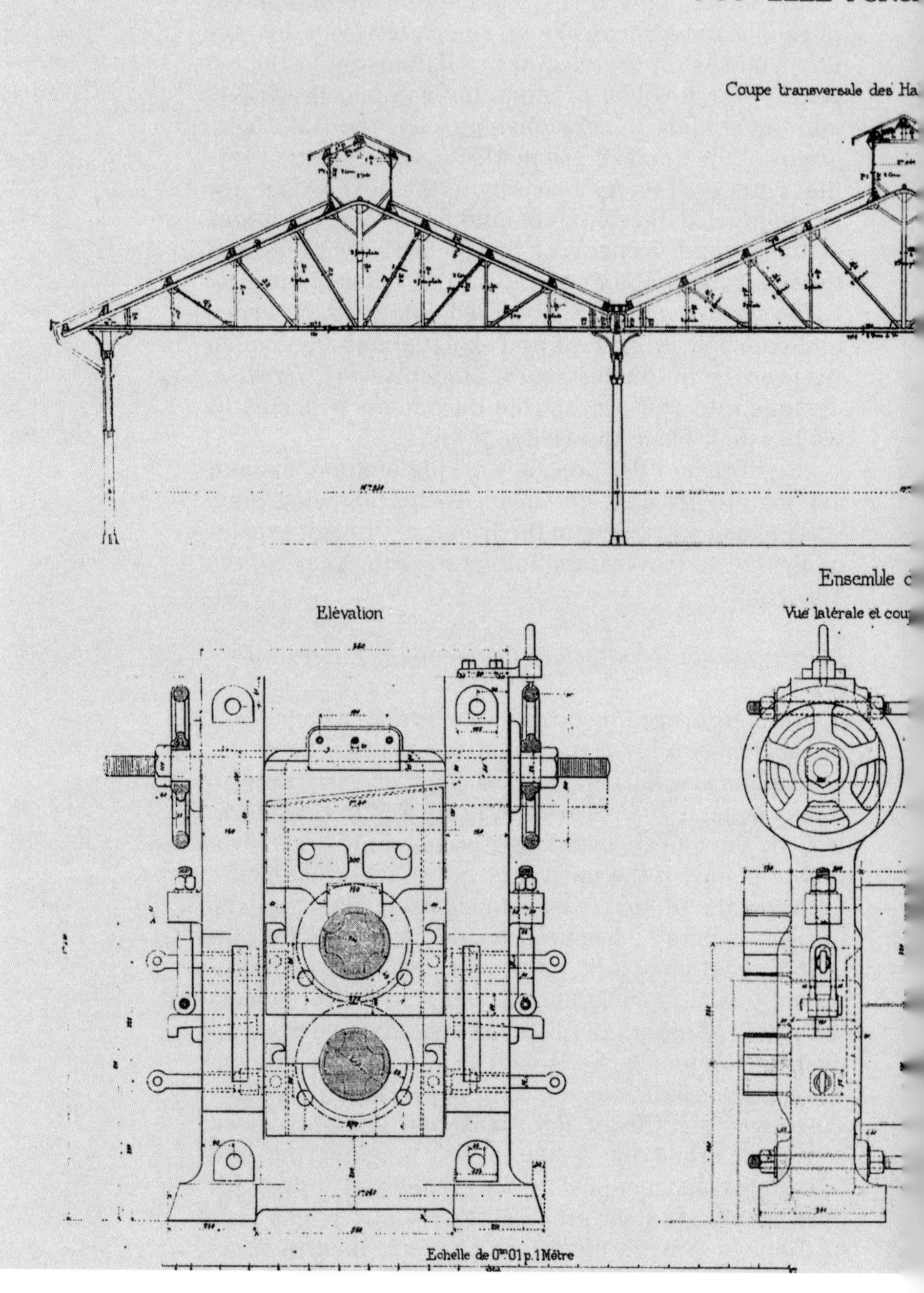
NOUVELLE FORG
Coupe transversale des Ha
Ensemble d
Elévation
Vue latérale et cou
Echelle de 0m.01 p. 1 Mètre

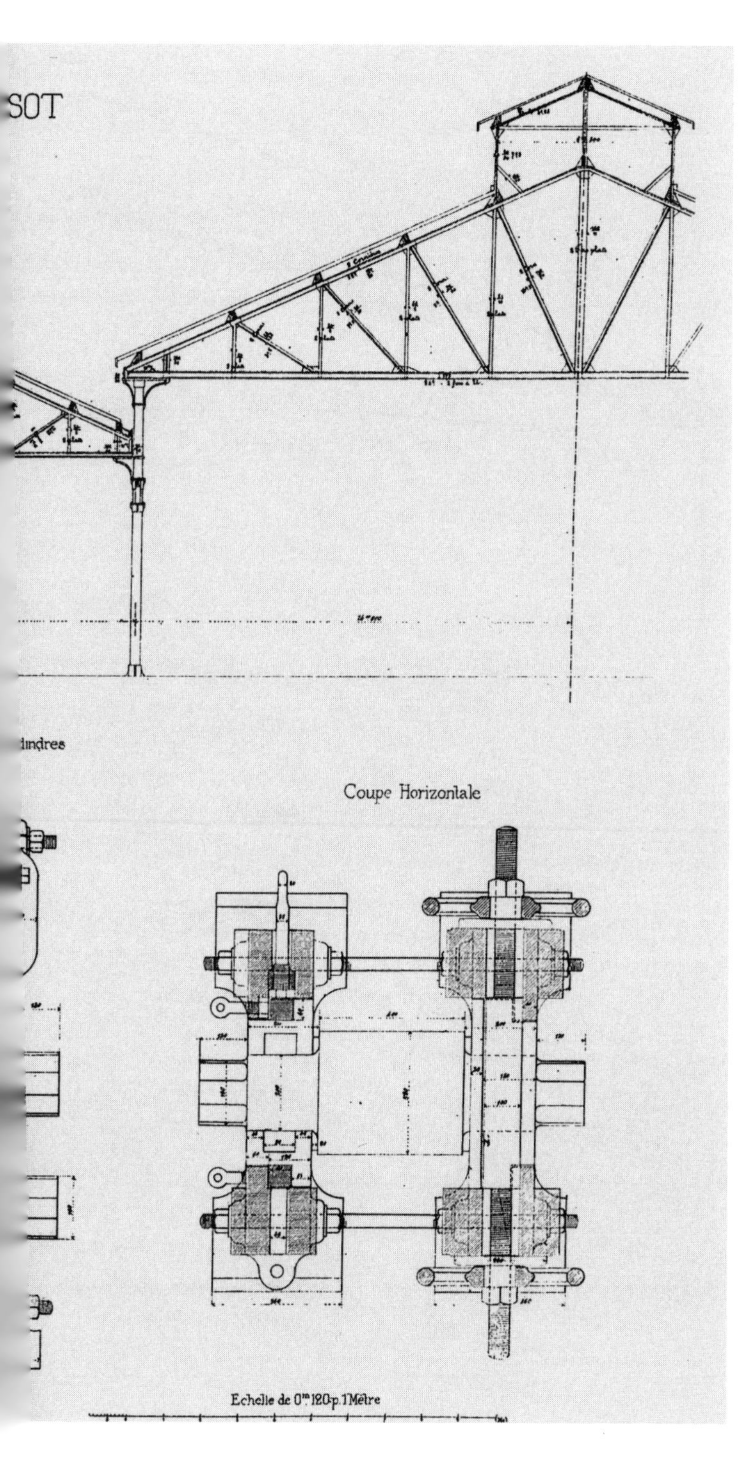

An example of vacation work executed by the student F. Campionnet, who analyzed the forge and metal workshop of the firm Creusot. (ECAM; Dossiers F. Hamon, Paris;©)

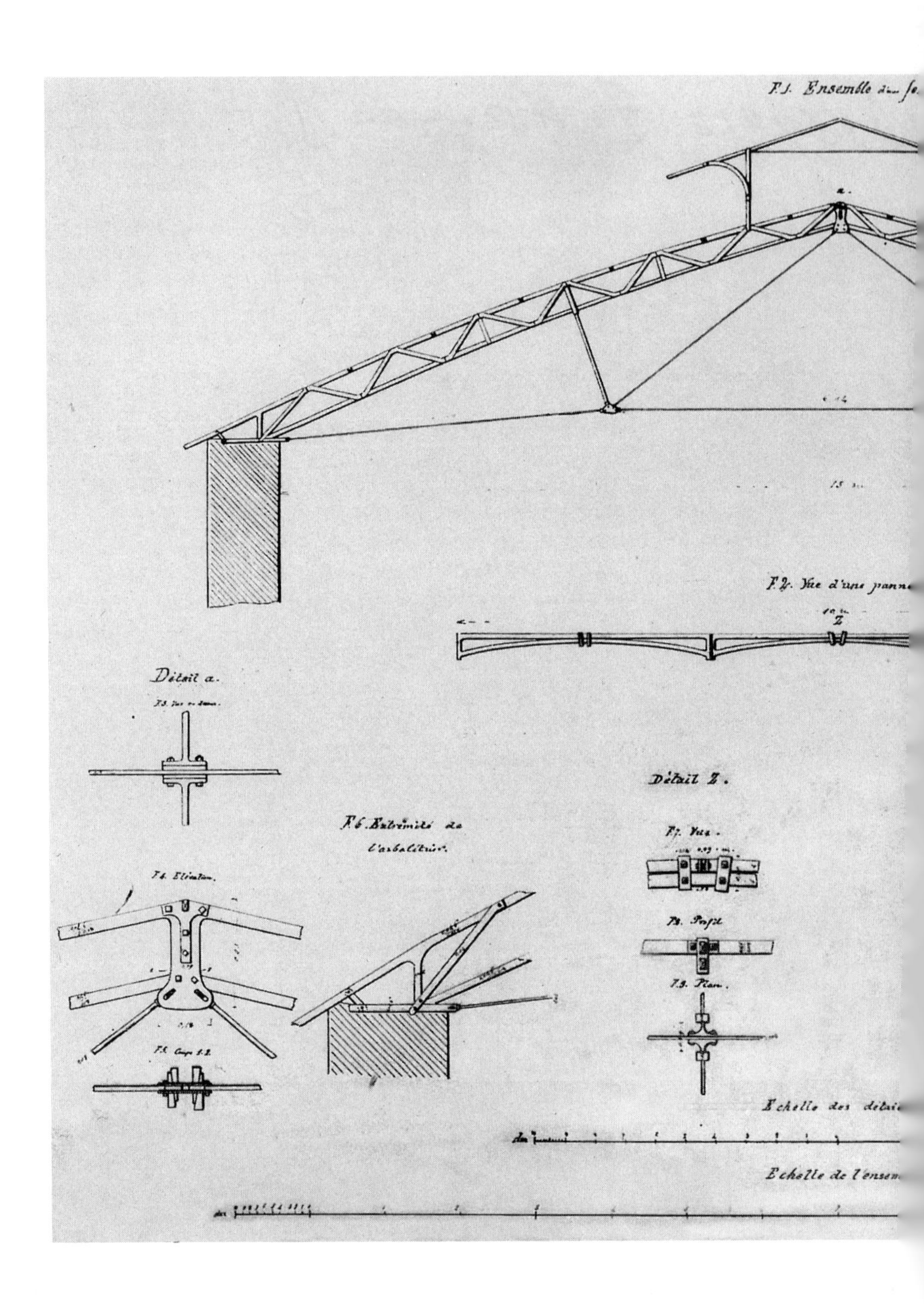
F.1. Ensemble d... fe...
F.2. Vue d'une panne
Détail a.
Détail Z.
F.6. Extrémité de l'arbalétrier.
F.4. Elévation.
F.7. Vue.
F.8. Profil
F.9. Plan.
Echelle des détails
Echelle de l'ensem...

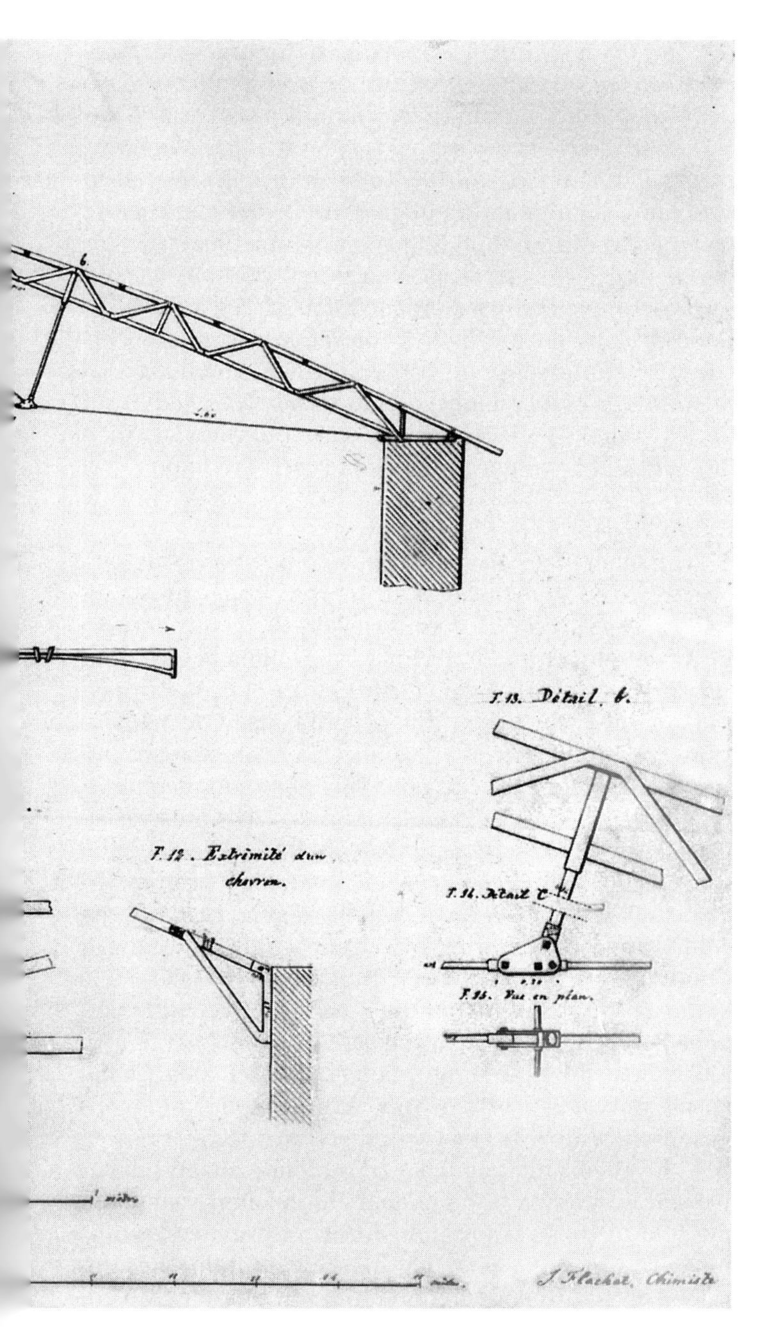

Final assignment in chemistry. Structural plan of a hall roof by J. Flachat (1850). (ECAM; Dossiers F. Hamon, Paris;©)

The École Centrale des Arts et Manufactures was the first educational institute of architectural engineering world-wide. In theory and in project work assigned to students the school responded to the modern construction activity of the 19th century, it referred to problems of engineering in its architecture classes and treated the nascent building and organizational assignments in project exercises such as railway stations and port facilities, factories, department stores and warehouses, covered markets, galleries and shopping arcades, greenhouses and winter gardens, not least also hospitals, social housing complexes, etc. The activity of entire student generations give proof of the sustained effect that this modern model of instruction had.

On the Discipline of Building

The division of architecture into composition (and decoration) and construction/structure, which began in the 19th century and manifested itself in the various forms of education for architects or engineers was not carried out by Mary in the École Centrale. He developed a teaching according to the school's basic principle – "La science industrielle est une" (a "sole science") – that treated both aspects equally and as complementary perspectives and working processes. The illustrative plates in his textbook demonstrate this, especially the example of the reconstructed/remodeled granary in Paris between 1809–1813, which was the result of the cooperation between architect and engineer. Sigfried Giedion remarks in his book "Building in France": "It is to our knowledge the first time that architect and engineer were no longer incorporated in one person."[43] Mary illustrated the Halle aux blés in two plates, one of space and structure, the other of technology and construction, which will be treated below.

[43] Trans. from Giedion, 1928, p.20, on illus.7.

The theoretical concept of a standardized education for architects and engineers found expression as a method in the tasks assigned during practice lessons and projects. Not least, the nascent building assignments of that time demanded that the relationship between composition and construction be redefined: According to Mary utility is to be equally considered as aesthetic agreeableness; structural requirements follow the building's function and spatial program, together they comprise the foundation for architecture, or for composition and form. The structural forms and solutions defined by function and material that were implemented in this model of industrial instruction and that are determined by circumstances of statics are the foundations upon which the *modern architect* rests. Modernism was built from this.

As a student of Durand, Mary played a key role in establishing the École Centrale; according to the chronicler Pothier, he was the actual founder and maker of the industrial model in its methodology of instruction.[44] This "school of thought", which formed whole generations of students, organized all the available knowledge of the time by developing a systematic theory and conceptual framework, applied this within a methodological model of teaching and learning and compiled exemplary didactic illustrative material. As a result, a professional ethic evolved which qualified the graduates as Ingénieurs centraux.

[44] Pothier, 1887, p.117.

Because the basic sciences were imparted to all students in the first half of their studies as a unified whole and because a specific specialized vocational field could then be chosen for in-depth study in the second half, the "Unité de doctrine" of interrelating all sciences with each other and with an industrial purpose remained the predominating underlying professional concept of the Centraux. It was a form of "industrial esprit de corps" that understood the object of industry to serve social progress and the betterment of the population's living conditions in accordance with the origins of the school in the circle around Saint-Simon. This spirit is also exhibited by Mary's successor, Émile Muller, in his commitment to social housing projects and in the achievements in infrastructure, the technical equipment and facilities of buildings, etc. on the part of uncountable graduates of the school.

Architects of the Beaux-Arts school such as Labrouste also adopted the industrially motivated principles of design and brought forth exceptional and unusually unique solutions, which can be seen in examples such as the Bibliothèque Ste Geneviève (1843–1850) or the Bibliothèque Nationale (1857–1867). Labrouste, however, did not receive academic acknowledgment for such work.[45] Victor Baltard, the builder of the Halles cen-

[45] Giedion, 1928, p.22f.

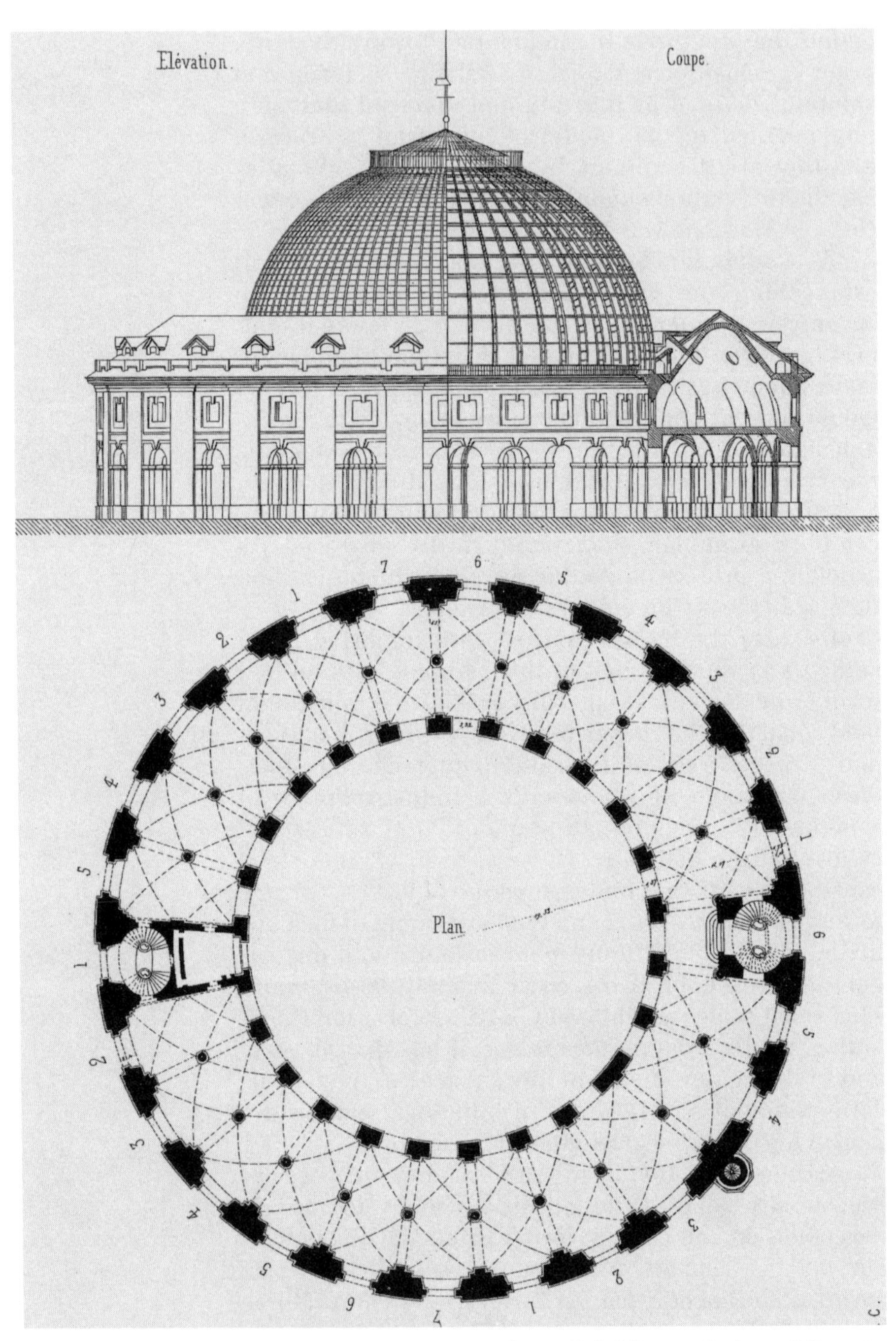

Halle aux blés, Paris 1811; Architect: F.-J. Bélanger, Engineer: F. Brunet. Parallel demonstration of architectonic and civil engineering in Mary's textbook. (Mary, 1852/53, Pl.X. and XI.)

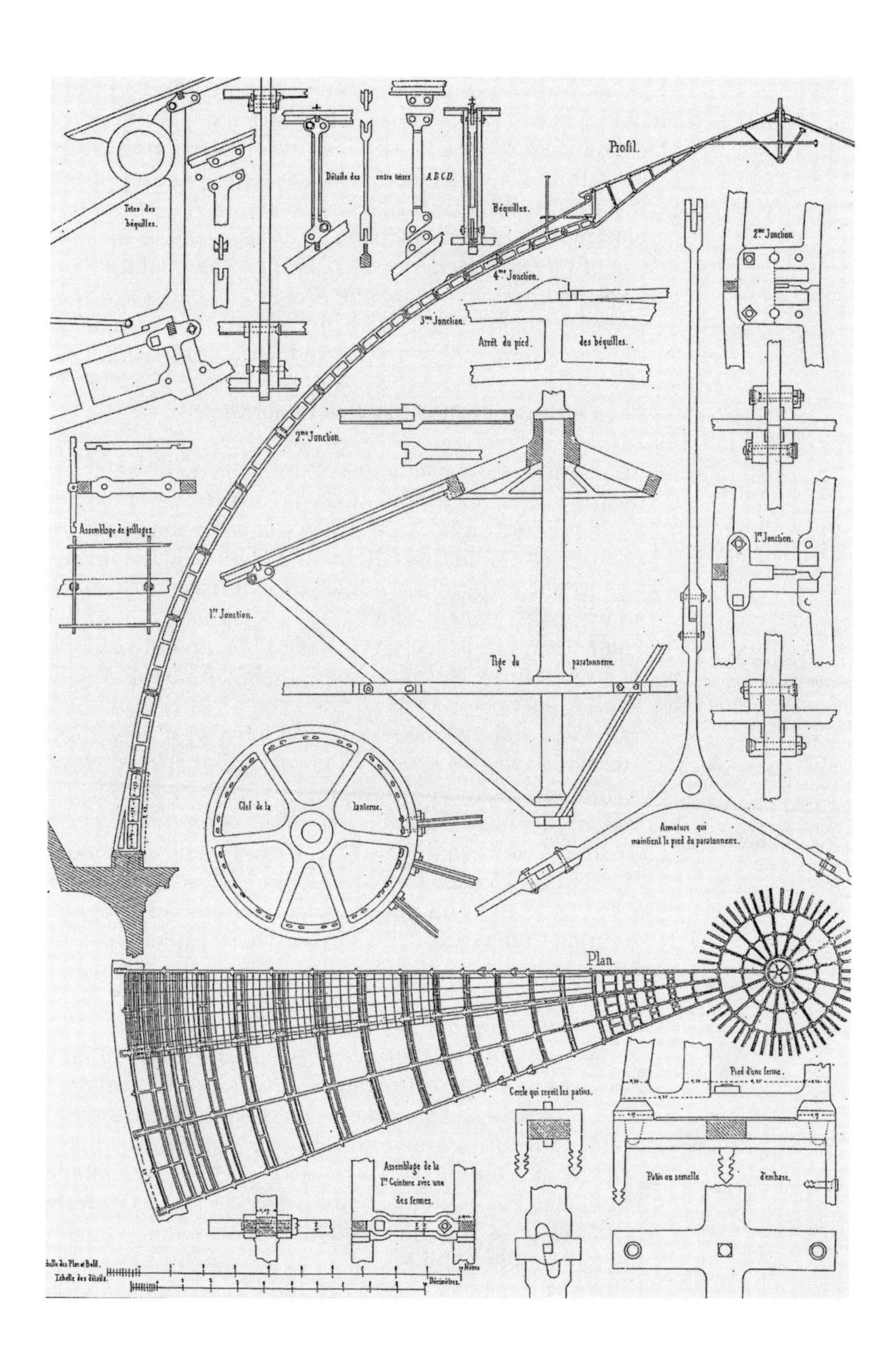
Profil.
Tetes des béquilles.
Détails des entre toises A B C D.
Béquilles.
2me Jonction.
4me Jonction.
3me Jonction.
Arrêt du pied des béquilles.
2me Jonction.
Assemblage de grillages.
1re Jonction.
1re Jonction.
Tige du paratonnerre.
Clef de la lanterne.
Armature qui maintient le pied du paratonnerre.
Plan.
Pied d'une ferme.
Cercle qui reçoit les patins.
Assemblage de la 1re Ceinture avec une des fermes.
Patin ou semelle d'embase.
Echelle des détails.
Décimètres.

trales (with Félix Callet) received his qualification from the École des Beaux-Arts with a "Grand-Prix d'Architecture" (1833); he was nevertheless trained by his father, who was the first architecture teacher at the École Polytechnique.[46] A further example of a "cross-border" or "Beaux-Arts engineer" is to be found in the person of Eugène Flachat, who as first president of the "Société des Ingénieurs Civils de France", founded in 1848, played a pioneering role in the political and technical representation of this nascent industrial profession.[47]

[46] Boussel (n.d.), p.8f.

[47] Jacomy, 1994, p.351ff.

The Centraux of the First Two Generations

Among the many graduates of the École Centrale from the time of its inception in 1829 up to its transfer to government control in 1857 a few Centraux stand out in the history of building culture and technology as inventors about whom little is known with respect to their educational background. This pertains to Camille Polonceau, Émile Trélat, Gustave Eiffel and the American William Le Baron Jenney. Included also is Émile Muller, successor to Mary as architecture professor, and Léon Édoux, pioneer in the history of elevators. The same era also produced three further specialists of technical building equipment: Léonce Thomas, D'Hamélincourt and Louis Ser. In the subsequent phase we find famous engineers of iron architecture: Armand Moisant and Victor Contamin, who still studied under Mary, and the concrete engineers Édmond Coignet, Maurice Dumas and J.T.Guéritte. These prominent "Ingénieurs centraux" will be introduced in the following.

Camille Polonceau

Polonceau (1813–1859) was the son of Antoine Rémy Polonceau, a Polytechnicien of the first hour (diploma 1796). It was characteristic for this era that sons of polytechnically trained fathers were sent to the modern and industrial oriented school.[48] Camille Polonceau completed his studies at the École Centrale in the year 1837 in the field of "Construction", meaning as engineer and architect at the same time. In his year of diploma he seems to have invented the prestressed support structure for covered halls.[49] Prominent applications of the "Polonceau Truss" also ap-

[48] A.R. Polonceau was a well-known general inspector of the Ponts et Chaussées, who together with the equally polytechnically trained engineers Coïc and Baduel cooperated on the construction of the strategic connecting roads between France and Italy (Simplon, Mont Cenis and Corniche); besides authoring a theory on bridge building, he also built iron bridges from the pioneering era such as the Pont du Carrousel; cf. Picon, 1992, p.363.

[49] Guillet, 1929, p.322.

[50] The general management of this exhibition was given to the Polytechnicien Frédéric Le Play, the structure in the oval exhibition pavilion was created by J.-B. Krantz and G. Eiffel;

peared, for instance, in Flachat's competition proposal for the market halls of Paris (1850), in the oval major pavilion of the Parisian World Fair 1867[50], in the covering-over of the tracks for the Gare du Nord in Paris (in its first construction by Reynaud 1842–1847 as well as in the second by von Hittorff and Couche 1861–1865), in the Gard d l'Ouest, where the Polonceau Truss was used by Flachat[51] or in Budapest's West Sta-

cf. Lemoine, 1986, p.220, and Giedion, 1928, p.40–43 and illus.37, p.44 (Section of the different layers of the hall).

[51] Lemoine, 1986, p.146.

COUPE TRANSVERSALE

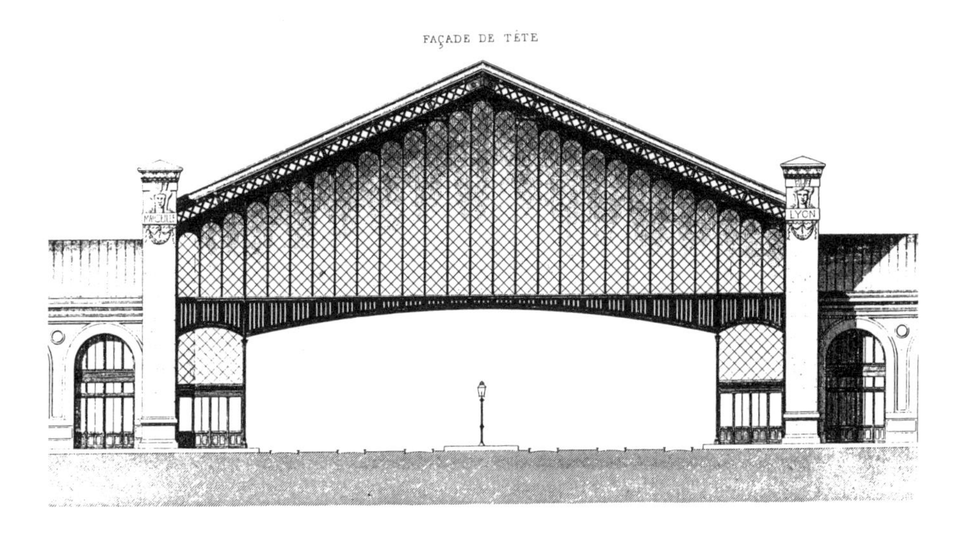

The use of the "Polonceau Binder" in the covering-over of the Station Lyon-Perrache (Architect: F.A. Cendrier, 1860). (Lemoine, 1986, p.145)

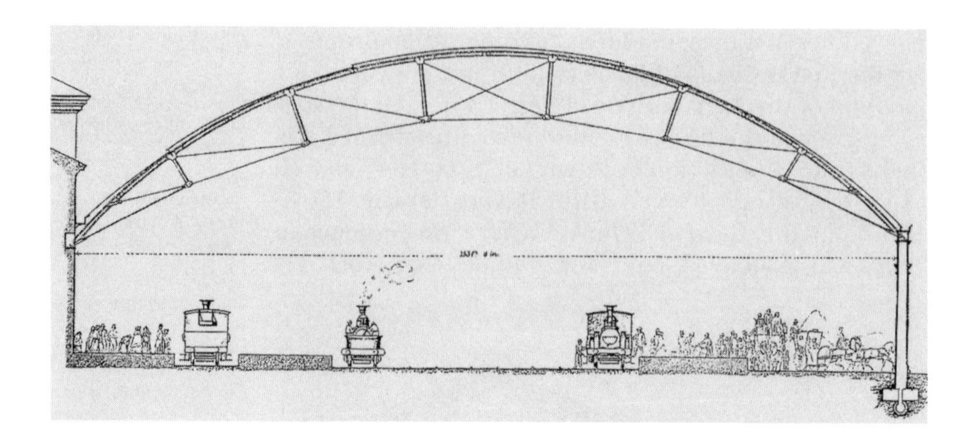

The roof over the Lime Street Station by Richard Turner, Liverpool 1849–1851. (Hix, 1974, p.126)

tion (1875/76) built by Eiffel, or the railway station hall Lyon-Perrache by A. Cendrier (1860). Turner developed an original interpretation of a flexed/semi-circular support structure in his Lime Street Station in Liverpool (1849–1851). [color plate 7]

This new form of support structure replaced the framework construction method in the formative years of the era of railways and exhibition halls and combined the overhead compression elements with tie rods. This not only saved material and weight but also considerably increased the efficiency or load-bearing capacity and spanning distances of the building structure. While spanning distances between 15 and 40 meters were first possible, the middle roof of the Gare d'Austerlitz spanned 52,55 meters.[52] However, such a typology of structure had already been in development in the school since 1835, which is illustrated by a construction plan executed by Delbar, Polonceau's student colleague. The traditional arcuated construction is, however, not yet eliminated but appears as a parallel and superfluous element: in breaking new ground the familiar was not yet abandoned.

Polonceau made his debut as one of the managing engineers of the railway line Paris-Versailles under Perdonnet, was subsequently director of the Alsace Railway and as of 1848 of the Compagnie Paris-Orléans. As railway engineer he improved the construction of locomotives and distinguished himself by important inventions in track construction, train wagon types, remote-controlled barriers, etc. and became particularly well-

[52] Architect: Louis Renaud, Engineer: Sévène; cf. Lemoine, 1986, p.72f., Marrey, 1989, p.32 (Marrey mentions 51.2 meters as the largest spanning distance).

[53] Guillet, 1929, p.320ff.

[54] Pothier, 1887, Synoptic Plate p.428f.

[55] Guillet, 1929, p.264.

One of the most perfect applications of the "Polonceau Truss": Gare d'Austerlitz with over 50-meter spanning distance (1865–1869); Architect: Reanaud, Engineer: Sévène. (Marrey, 1989, p.33)

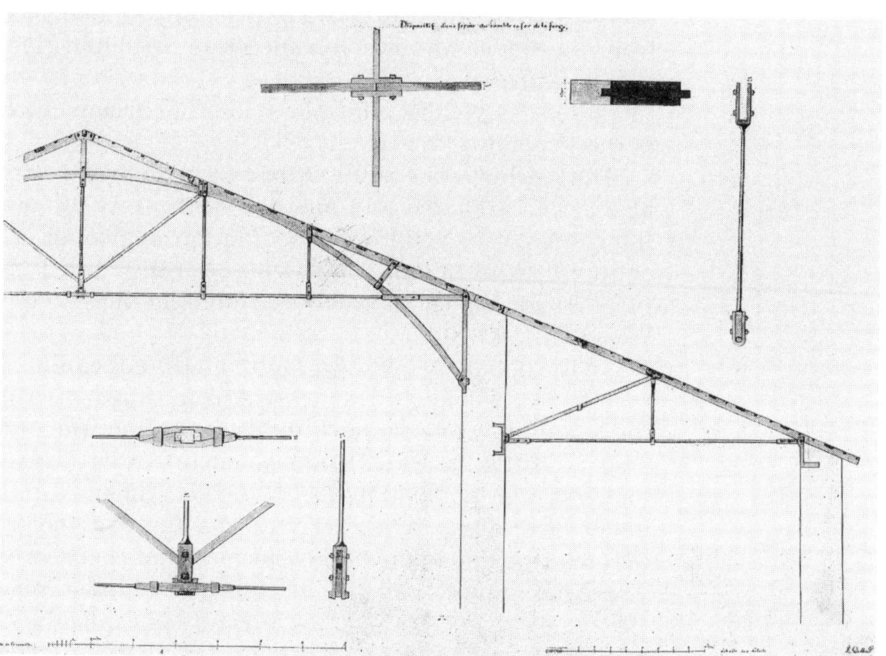

Final work (diploma) by Delbar in metal construction at the École Centrale (1835): A prototype of the "Polonceau Truss". (ECAM; Dossiers F. Hamon, Paris;©)

known for developing a turntable for locomotive workshops.[53] During the year-course 1851/52 Polonceau lectured in mechanical engineering at the École Centrale[54] and presided at the Société des Ingénieurs Civils de France in 1856.[55] At the same time Polonceau taught at the school, Eiffel began his studies there.

Émile Trélat

Trélat (1821–1907) received his diploma from the École Centrale in 1840 in metallurgy.[56] In 1854 he became professor of "Constructions civiles", which was newly established in that year at the Conservatoire des Arts et Métiers.[57] Trélat organized his construction course in two parts: the science of structure and material and the planning and realization of buildings. In his historical and critical studies upon which he based his classes, he emphasized that the main purpose of architecture is to give the idea and concept derived from the building program and its function a most perfect expression, and that this has always characterized the buildings representative of every epoch. Trélat introduced technical building equipment, the "science of hygiene", as a subject in his class. In 1894, after 40 years of teaching, he resigned from the Conservatoire. His successor was Jules Pillet, a Polytechnicien, engineer at the Ponts et Chaussées as well as a former student of architecture at the École des Beaux-Arts.

[56] The report follows Guillet, 1929, Seitz, 1994, and 1995, as well as Claudine Fontanon, 1994.

[57] In this field Jean Prouvé was also to teach 100 years later (1958–1970).

During his work at the Conservatoire Trélat was also acting architect and member of the Jury for the International Exhibition of 1855 in Paris, also an architect for the Italian railway (1856–1860) and jury member for the competition surrounding the London World Fair of 1862.

In the debate following a report on the education of architects at the École des Beaux-Arts in 1863 between the traditional "classicists" (or "Romans") on the one hand and the modern "Gothics" on the other, Trélat took a similar position as Viollet-le-Duc. The priorities the École des Beaux-Arts set in the master-oriented art of rendering and in the highly coveted Rome Prize provided the impetus for and helped to establish a new "free and private school of architecture of France".[58]

[58] Seitz, 1995, p.19–26.

In this context Trélat therefore founded the École Spéciale d'Architecture in 1865 with the goal to train skilfull and able design engineers and competent professionals. His new school was supported in the beginning by Viollet-le-Duc and Labrouste.[59] According to Trélat the "architect engineer" was to combine engineering, material-technological and economic skills with those of an artist, and be capable of giving his

[59] Viollet-le-Duc tried at that same time (1864) to teach at the École des Beaux-Arts, resigned, however, already after one month!

thoughts and ideas a precise and intelligible form. At the same time, according to Trélat the architect was to also perform his profession in the service of social welfare, and not only to concern himself with monumental and government facilities but also to address the numerous small, local and unspectacular assignments of daily life.[60]

Trélat thus reconnected with the tradition of the Enlightenment, which had already found expression in the activity of the École Polytechnique and the École Centrale des Arts et Manufactures, and, in contrast to that of the École des Beaux-Arts and from a historical perspective, provided his school with a liberal scope. The École Spéciale d'Architecture, ESA, was first titled École Centrale d'Architecture in reference to Trélat's background and to a few other promoters and teachers of this new institution. On 30 June 1865 it received government authorization and on 10 November 1865 it began its educational activity.[61]

Trélat embodied the Ingénieur centrale type based on the professional social and educational ethic nurtured by the École Centrale. He also represented this ingenious esprit de corps in his function as president of the Société des Ingénieurs Civils de France in his year of office in 1882.

Émile Muller

Muller (1823–1889) was born in Alsace and completed his studies at the École Centrale in 1844 as "Constructeur", as architect-engineer.[62] In 1864 he was chosen as Mary's successor for teaching the theory of architecture and structure/construction, were he remained until his death. Before the German annexation of Alsace in 1870 Muller worked in there as an architect and industrialist, acquired a reputation for building public sanitary systems such as baths and washing facilities and e.g. for building workers' housing in the suburbs of Mulhouse. In 1854 he founded a ceramic factory in Ivry.

Muller's social commitment as an architect and businessman in Alsace resulted in addressing questions of social economy and subsequently in the establishment of various interest groups such as the "Société Industrielle de Mulhouse" (1867), which he

[60] Seitz, 1995, p.27ff.

[61] Some of the most famous graduates of this school were Henri Prost (diploma 1892), R. Mallet-Stevens (1906), Jean Ginsberg (1928), Pierre Vago (1933) and others. Behind the recent presidents Paul Virillo figures.

[62] Guillet, 1929; Kalaora and Savoye in: Grelon and Stück, 1994, p.266f.

used as models for similar associations in Paris that he also called to life: a "Society for Sharing Profits" (Société pour la participation aux bénéfices) or a "Society for the Protection of Apprentices" (Société pour la protection des apprentis). In continuing the spirit of the Enlightenment he sought to increase the welfare and the quality of life of disadvantaged worker families, especially with respect to their basic hygiene by building sanitary facilities in living quarters or housing complexes. With his summary work on "Workers' Housing of all Countries" (Les Habitations ouvrières en tous pays) published in 1879 Muller made a major contribution to research in this field, which was taken up in 1889 on the occasion of the first "International Congress for Inexpensive Housing", in which Muller also participated, leading to the founding of the "French Society for Housing Construction" (Société française pour les HBM).

Muller made this an issue at the École Centrale after he succeeded Mary's long tenure as architecture professor in 1864. As a teacher of architecture Muller endeavored in his course "Constructions civiles" to address the meanwhile unlimited possibilities of industrial production in class and – just as the school's founders before him – in his application and practice-oriented exercises and projects. Of the project assignments given to his students a few have survived from the year 1875. The final or diploma project of that year was titled "Building Atelier-Apartments" (Immeuble industriel). What was required was a large housing block on an inner-city street combined with a series of workshops or ateliers for manufacturing that were connected to the respective apartments. Each apartment was to have a WC and kitchen installed, the entire complex to be provided with central heating and washing facilities.

This astonishingly early *integral approach* in Muller's construction course shows the École Centrale and its faculty's unbroken position of leadership in the industrial era. Moreover, Muller also paved the way for a central issue that was to concern modernism between the wars. Muller took over the presidency of the Société des Ingénieurs Civils de France in 1872 and in 1880 established its organ "Le Génie Civil".

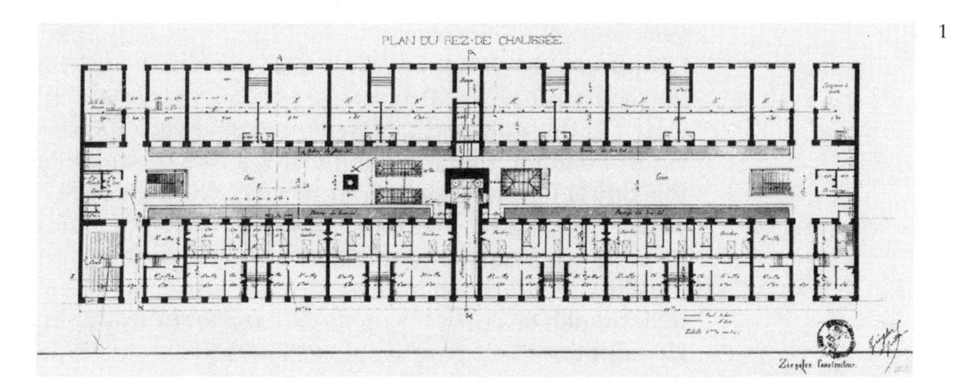

1

2

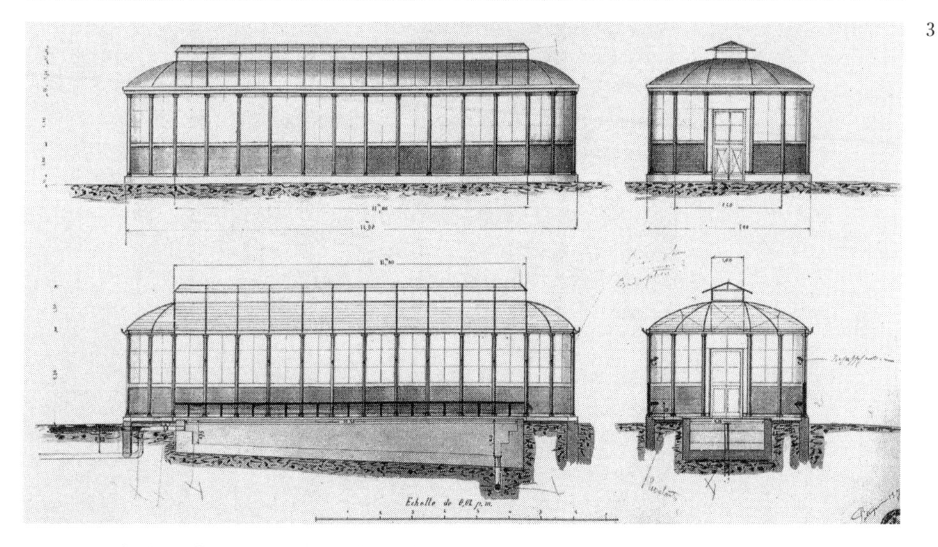

3

The entire facility of an "Immeuble industriel" Illus.1: Apartments with kitchen and WC facing the street; in the block facing the yard ateliers and workshops, and at the corners shops and a school. The blocks are connected at the front by passageways, which contain the entrances and concierge-lodge on the ground floor; the central unit allows the "house technician" access to the central heating system. The middle yard carries light and air to the work-places and allows cross-ventilation in the apartments (Student Ziegler). Illus.2 shows the delineation between the representative street and industrial front facade made of iron and glass, which provides maximal lighting (Student: Hardon). There are bath pavilions in the yard, which can be seen in more detail in illus.3 (Student: Petrequin). (ECAM; Dossiers F. Hamon, Paris; ©)

Léon Édoux

Édoux (1827–1910) received his diploma from the École Centrale in 1850 as mechanical engineer ("Mécanicien"). He is one of the pioneers of the hydraulic elevator; he titled his construction "Ascenseur", which to this day is the common word in French for a lift or elevator.[63] Because the steam engine had reached the limits of its capacities and the electric motor was not yet available, Édoux decided to exploit water pressure, which could be tapped from Paris' canalization system. He built the first operable Ascenseur already in 1864. While the American Elisha Otis had already patented

[63] Guillet, 1929, Dubin, 1981.

First hydraulic person lift by Édoux in the exhibition pavilion of 1867. (Simmen, 1984, p.61)

his security system for an elevator in 1853 and subsequently developed a few goods lifts for skyscrapers in New York, the first of his "Otis" person lifts were not installed until 1870 in New York's Equitable Building; this then triggered the entire wave of high-rise construction. As of 1867 the first goods elevators were put into operation in Great Britain.[64] Édoux' invention of 1864 was therefor a world-premiere. Its first attractive implementation took place in the pavilion of the International Exhibition in Paris in 1867.

[64] Peters, 1987.

Édoux' teachers included Jean-Baptiste Bélanger (mechanical science, applied mechanics, hydraulics), Léonce Thomas (steam engines), Walther de St.-Ange (mechanical engineering), Perdonnet (railway engineering), Ferry (metallurgy) and Mary (architectural engineering). As inventors and pioneers they were all active in various fields and inspired their students with the spirit of invention and scientific research. Édoux subsequently constructed a 60–meter-high person elevator with 80 seats for the Palais du Trocadéro tower at the Paris Exposition of 1878. On the occasion of the World Exhibition of 1889 in Paris he installed a double lift for 65 persons from the second to the third floor of the Eiffel Tower, whereas Otis built the direct diagonal lift from the ground floor to the second platform (the bottom diagonal lift to the first platform was installed by the firm Roux, Combaluzier et Lepape).

Édoux lift in the Eiffel Tower (2nd to 3rd platform) with two cabins for 65 persons each; they accomplished 7 trips an hour. While the ascending cabin was hydraulically powered from the intermediate floor, the descending cabin acted as a counterbalance. Illustration: Changing cabins at a stop. (Simmen, 1984, p.238)

Gustave Eiffel

Eiffel (1832–1923) was to choose an appropriate career according to his father's wish, who was an officer; in 1852 he tried to pass the École Polytechnique's entrance exams. He failed in mathematics and, as did many other candidates, decided to study at the École Centrale. In the same year he passed the entrance exams. François Poncetton, engineer and Eiffel's biographer, wrote that many young students knew that this engineering school would provide them access to modern industry and thus opening up the most hopeful and at the same time adventurous future: "While the Polytechnique assured a career, the Centrale promised industrial competition."[65]

[65] Poncetton, 1939, p.68.

Eiffel fought to improve his drawing ability during his studies, which was strongly required in all subjects at the school; he passed his intermediate exams and decided in the second half of his studies to specialize in chemistry. The reason for this was that his uncle, Jean-Baptiste Mollerat, was a chemist who owned a distillery and had named Eiffel as his successor. However, Mollerat died while Eiffel was still studying and the firm was forced to close due to the lack of an interim manager. Eiffel still completed his studies as an engineer-chemist in August 1855. His final project is stored in the archives of the École Centrale: the arrangement of a large agricultural complex with living quarters, stalls and a coach house. The complex's rational composition, based on Durand's "papier quadrillé", and its standardized structural construction reveal his interest in architecture and the spirit of this future engineer of metal. His teachers included Ferry, Perdonnet and Mary.

As a student Eiffel lived in a small furnished room in a guest-house near to the École Centrale. He lived modestly, his mother provided him with whatever was necessary and desired, she also paid his lodging. As was usual for the Centraux, one studied until deep into the night, completed one's homework and finished project work. Sundays were reserved for outings. Eiffel often went to the ocean, frequently to Dunkerque. On free afternoons he visited factories and workshops or he took long walks. During vacation the Centraux underwent study trips as well as completed vacation work. Eiffel's graduation trip took him to Switzerland.

In 1856 he sought employment as "Ingénieur des Arts et Manufactures" in Paris. In an industrial area of the city he came across a machine and tool factory, the owner of which, Charles Nepveu, hired him immediately. He remained at this firm for a while as an engineering consultant, worked for a short time in between for the Compagnie des Chemins de fer de l'Ouest, where other Centraux were employed and which was more lucrative employment; and in 1857 he returned to Nepveu who engaged him as project manager for an important railway bridge being constructed for the Compagnie du Midi in Bordeaux. The bridge in Bordeaux was Eiffel's first independent work and soon became renowned for its originality; the construction lasted between 1858 and 1860, the bridge was 500 meters long and was made up of a system of latticed girders resting on masonry pillars which were underpinned by means of a novel compressed-air method. Between 1862 and 1864 Eiffel worked as director of the firm Pauwel & Cie. in Clichy and was subsequently engaged as a free-lance engineering consultant.

In 1866 Eiffel established his own construction workshop in Levallois-Perret near Paris with the help of his father-in-law. Eiffel's business built many bridges, such as the Sioule Viaduct (1867–1869), the Douro Double Bridge in Porto (Northern Portugal, 1875–1877), the Garabit Viaduct (1879/80) and many others; in addition, the firm built factories, churches, the West Station in Pest (Budapest, 1875–1877), and so on.

The department store Au bon Marché in Paris, specifically the second phase in metal, was a product of the initial cooperation between the architect L.-C. Boileau with the metal construction firm Armand Moisant and then with Eiffel (1879/80).[66] In the same period the Eiffel firm built the Paris International Exposition's large glass porch in 1878.[67] Eiffel also planned and designed bridges that were capable of being dismantled and rotated, he planned a twelve-kilometer-long city and circular belt around Paris, a double-conduit tunnel under the English channel and experimented in aerodynamics towards the end of his life, producing an airplane prototype.

[66] Moisant was also a Centraux graduate and received his diploma four years after Eiffel.

[67] The Statue of Liberty, as France's gift to the United States, was also constructed in Eiffel's workshops (1881, inaugurated on 28 October 1886).

Plan des batiments a

Vacherie

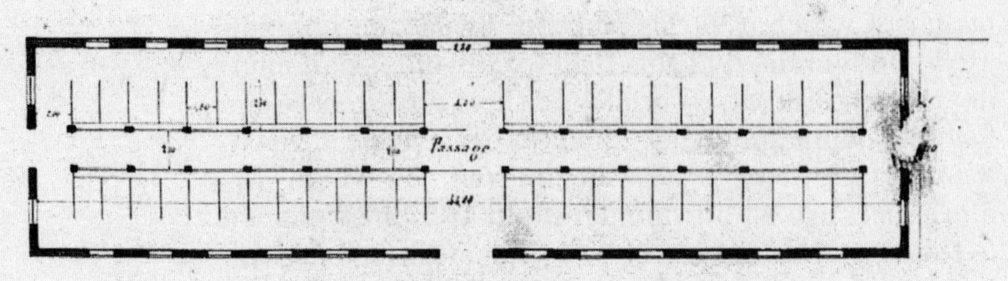

Ecurie

Hangar et Ga

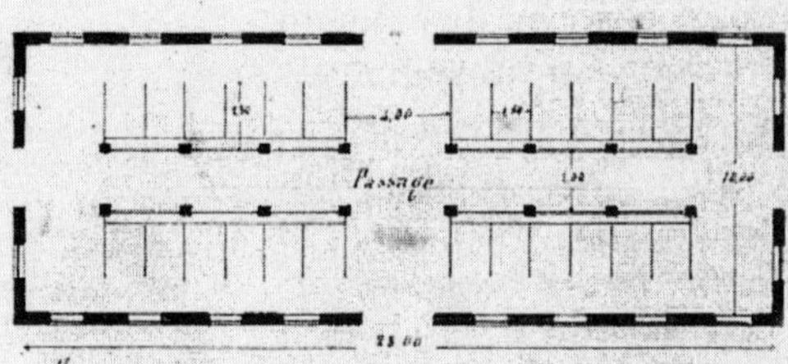

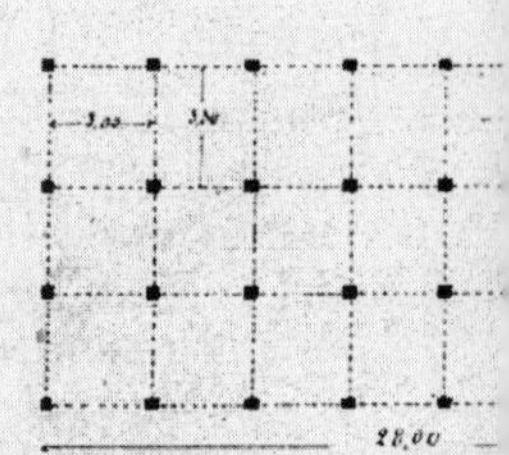

Bergerie

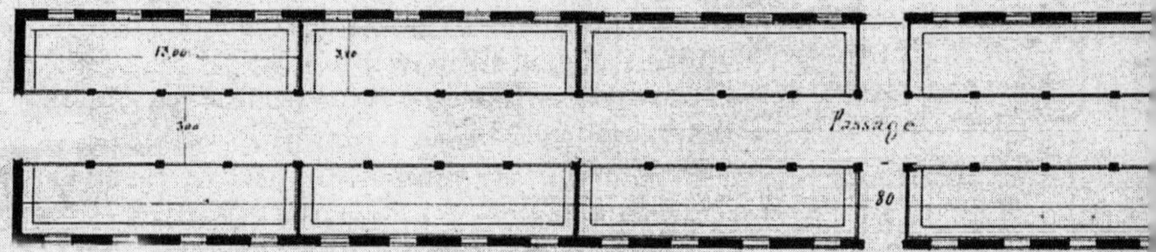

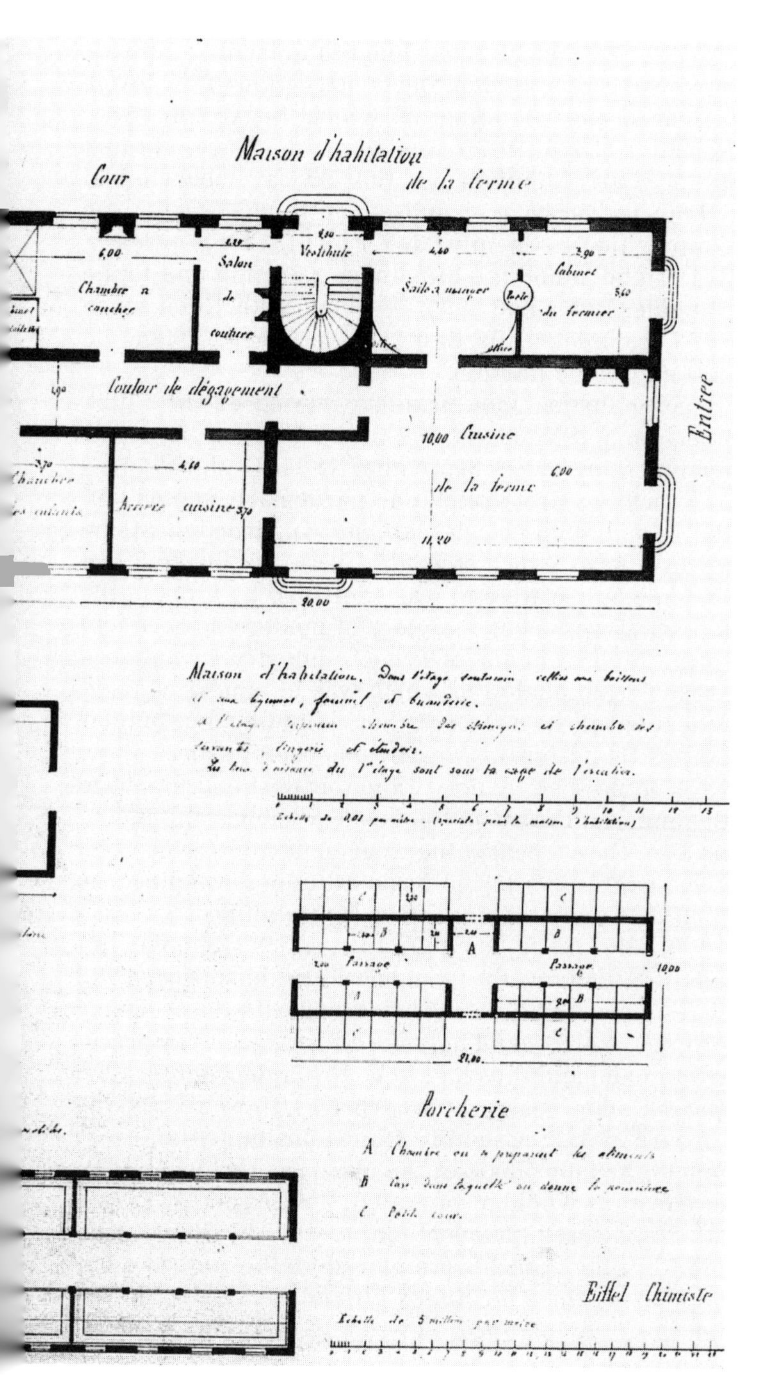

Eiffel's final graduating project in the field of "Chimie industrielle" (1855) at the École Centrale; arrangement of buildings for an agricultural enterprise. (ECAM; Dossiers F. Hamon, Paris;©)

Maurice Koechlin was one of the leading individuals who cooperated on the construction of the tower in Paris, the planning of which began in 1884. Koechlin (1856–1946) was born in Alsace, studied under Carl Culmann at the Swiss Polytechnical School in Zurich and subsequently worked for the Compagnie des Chemins de fer de l'Est for two years, in 1879 he joined the Eiffel firm. Koechlin and Émile Nouguier, who had been working with Eiffel since 1875, completed the first sketches for a "Pylône de 300 mêtres de hauteur".

After further modifications carried out by Eiffel, Koechlin and Nouguier, the firm submitted the tower for patenting on 18 September 1884. Of the changes to Koechlin's first sketch one particularly stands out, the added semi-circular arched construction at the tower's base which was the work of the architect Stephen Sauvestre and was explained by Eiffel in the following way: "For the glory of modern science and to the highest honor of French industry I wanted to erect a triumphal arch which was to appear as magnificent as those our forefathers built in honor of their heroes. Is not in this an obvious useful aspect of my work to be seen, which, I hope, is destined to add to the honor of France in the fields of science and industry?"[68] Here speaks the Ingénieur industriel, who gives expression to the basic attitude maintained by the École Centrale since its very beginning.

[68] Trans. and cited according to Von Büren, 1988, p.7.

Besides the finished product, which still today fulfills its function, the engineers and businessmen of that time were primarily concerned with the assembly process, which could not take advantage of any precedent model and thus had to be invented from scratch. Here a step-by-step approach was chosen involving upward sliding assembly cranes positioned on every one of the four pillars at the place from which the later elevators would be operated. Besides access cranes steam-powered lifts were employed. After the second platform the cranes were placed on the central pillar supporting the later Édoux lift and rose upwards. After the third platform only simple building winches were used.[69]

[69] After the description of Eiffel's assembly procedure, in Von Büren, 1988, p.56f.

The 300–meter tower embodies the collaboration between the industrial and polytechnical schools of

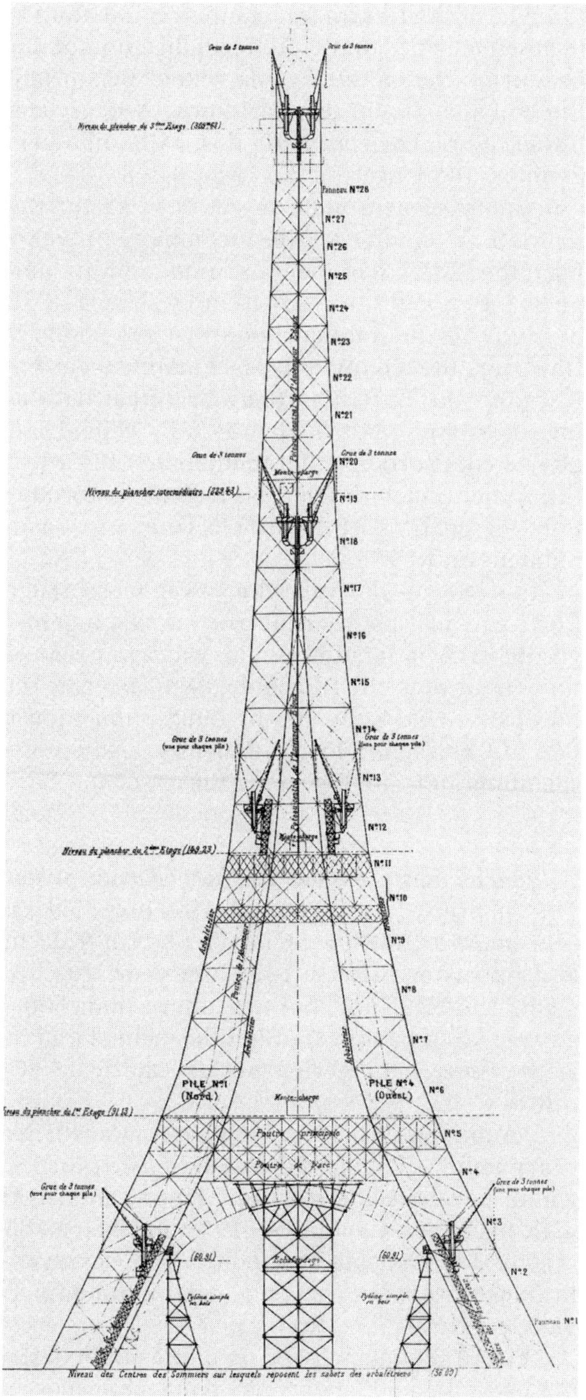

Assembly plan for the 300–meter tower according to Eiffel. (Von Büren, 1988, p.56)

thought of the Centralian chemist Eiffel and the Polytechnicien mathematician Koechlin from the Culmann School in Zurich; since this school was oriented on the Polytechnikum in Karlsruhe, which itself was based on the Parisian École Centrale, the circle of Ingénieurs industriels is thus completed.

After the tower in Paris was constructed the "Panama affair" troubled Eiffel's company. It was to support the Compagnie de Panama and in particular push forward the construction of sluices so that the deadline for the inauguration in 1890 would be met. However, the Compagnie de Panama became insolvent and the Eiffel Company braced it with a major investment inflow, which was only repaid later and after a court order. Eiffel then shifted his attention to experimenting with and researching in aerodynamics until his death in 1923. Eiffel's company was not liquidated until 1975.[70]

[70] On the construction of the Panama Canal, cf. Peters, 1996, p.295–345.

In 1889, in the year the tower was inaugurated, Eiffel was president of the Société des Ingénieurs Civils de France. In addition, he was awarded prizes by the "Institut", the Société d'Encouragement pour l'Industrie Nationale and by the Smithsonian Institution in Washington D.C., and was president of the Association amicale d'Anciens Élèves of the École Centrale.[71]

[71] Guillet, 1929, p.252, 264 and 312f.

William Le Baron Jenney and the "Chicago School"

The American Jenney (1832–1907) studied as of the year-course 1852/53 at the École Centrale in Paris and graduated there in 1856, one year after his fellow student Eiffel. Until that time more than 600 foreign students had passed through this school and in 1865, at the time of the Civil War, 53 Americans were enrolled.[72]

[72] Turak, 1986, p.28; the following biographical notes originate from the detailed elucidation by Turak and the Autobiography: Jenney, 1906/07.

Jenney was born in Fairhaven (Mass.). As a 17-year-old he left his hometown for San Francisco, from where he endeavored to sail around South America with his friend Captain Scott. In February 1850 they returned to their starting-point. Jenney stayed in San Francisco until the city's wooden buildings were completely destroyed by fire in just three days starting on 18 May of that same year. Just one stone building remained intact. After this shocking experience Jenney continued to sail around the world, boarded a clipper headed for the Sandwich Islands, visited Manila and viewed the unique stave and lake buildings there, went on to China and Java, sailed around the Cape of Good Hope, took bearing for St. Helena and reached New York in 1851.

In the same year Jenney began his professional training at the Lawrence Scientific School, the engineering department of Harvard University in Boston. In a discussion with Cuban students he heard of the Parisian École Centrale des Arts et Manufactures, which was superior to Harvard. He then wanted to apply to this world-renowned school and learned French in order to pass the exams. He continued to study at Lawrence Scientific school for just two more years.[73] Jenney then took part in a three-month preparatory course in Paris in 1852, improved his French and finally passed the École Centrale's entrance exam in September of that year. He passed the first year-course (1852/53) brilliantly and was rated among the first 100 students to be accepted to the second year (1853/54). After the third year-course, in which he decided to take "Construction", meaning architecture or civil and public engineering, Jenney received his diploma as Ingénieur industriel (1856).

[73] The reference written by this teacher at Harvard, Prof. Henry L. Eustis, is dated from 11 June 1853.

Turak, Jenney's biographer, describes the course of his studies in detail and particularly points out Jenney's impressive list of projects for just the second year alone: a school house and a railway station complex, a wooden bridge and a tunnel, two roads and a water-supply facility, a public bath and a heating system for a hospital, study projects for a smelting oven and an elevator system, furthermore, experiments on material resistance, topographical analyses and presentations in the course on landscape planning, and more.[74]

[74] Turak, 1986, p.32.

Besides the importance of Mary's architecture class as a whole, it was those lessons on walls and masonry that determined Jenney's conception and treatment of the issue in the context of the "Chicago School" he founded after the major fire of 1871. The distribution of loads, of compression and tension elements and their visible manifestation in an interplay of supporting and non-supporting, closed and open structural components represents an early form of functionalism in modern architecture. The principle of "Pan de fer" or "curtain wall" was a standard element in Mary's construction course and corresponded to the state of development of architectural engineering of that time, which is illustrated in the compendiums published for instance by Charles Eck or Léonce Reynaud.[75] Such books were also used at the École Centrale, which has been documented by Turak.[76]

[75] Steiner, 1984, Pl.61, and Turak, 1986, Fig.3, 4 and 5; Eck, 1841.

[76] Turak, 1986, p.41.

The roof-constructions taught by Mary entailed pre-stressed inverted T-girders (standardized steel girders) resting on a wall or on cast-iron supports; placed between the flanges a vaulted ceiling with hollow units made of fired clay was laid, which was additionally held together by horizontal tie rods. This standard construction, which originated from the cotton mills of England's Midlands and in the process of industrialization also appeared in France, is to be found in many of the textbooks of the time, for example in that of Eck and also in those of Mary; this has already been mentioned. Mary's student Jenney further developed and adapted this principle to American conditions and invented the "Chicago frame", a typical construction of the "Chicago School of Architecture", as well as a "typology of joints".

Jenney's architecture teacher Mary nurtured a consciousness for technical building facilities astonishingly enough early on in his course. This not only found expression in his classes on factory construction but also in his teaching on urban rental housing and French rural building. In domestic construction Mary demanded that enough light, fresh air and heating, a well-placed kitchen with enough pantry space, baths and toilets be included in the design. An key quality in modern housing plans and domestic hygiene was, according to Mary, the correct orientation of the various functions; for example, the kitchen, because of the perishable goods being used, was to be placed in the north. Spatial features, climatic comforts and easy use were subjects treated in the lectures and in the study projects. This experience later also influenced Jenney's housing construction activity in America and finds expression in his only book, “Principles and Practice of Architecture” published in 1869.[77]

[77] Jenney, 1869.

After graduating in 1856 Jenney traveled to Genoa over the Alps, boarded a ship for Mexico, where he found his first employment as an engineer. After three months he was sent by the same firm to New Orleans and later to Philadelphia. Engaged by an American railway company he was to undertake research in Paris and Europe and began to plan his trip, but was interrupted by the American Civil War (1861–65). He subsequently belonged to General Sherman’s staff as a cartographer, and after 1865 drew his entire deployments of the Civil War in a series of maps; they were then printed and published under the new president, General Grant.

In 1866 Jenney again traveled to New York, where he was offered a chair in engineering at the New Brunswick College in New Jersey, but which he declined in favor of a job as manager of Humboldt Oil and as engineer for McKean County Coal Mines.

He finally moved to Chicago and initially became a partner in the architecture firm of Sanford E. Loring. After the “Great Fire” of October 1871 in which a great part of the city was destroyed, the architects of Chicago had all very much to do. At this time he set up his own office. Jenney’s first building, the Natural History Museum of the University of Michigan in Ann Arbor, was constructed in 1875 in the best Beaux-Arts style, however incorporating the most modern British and French iron architectural engineering techniques learned from his teacher Mary at the École Centrale. In 1876 the University of Michigan appointed him professor of architecture. There he remained until 1880. In 1883 he held a series of lectures on the history of architecture at the University of Chicago and published them in 13 issues titled “Architecture” in the journal “The Inland Architect and Builder”.[78]

[78] Jenney, 1883/84.

As of 1879 Jenney primarily worked on commercial building assignments in the “Loop” business section of Chicago. He built the famous First Leiter Building (1879), the Home Insurance Building (1885) and many other office buildings; he is also to be considered the actual founder of the Chicago School of Architecture.

W.L.B. Jenney: First Leiter Building (1879), forerunner of the “skyscraper”. (Condit, 1964, illus. 40)

Two of Jenney’s most important colleagues and assistants in his architectural firm were William Holabird and Martin Roche. Holabird entered the firm as a 21–year-old in 1875 and Roche already in 1872. In 1880, together with Ossian C. Simonds, Holabird opened his own office; in 1881 Roche joined him and as of 1883 the firm was called Holabird and Roche. The firm operated and continued to be successful until 1927.[79] Louis Sullivan also spent a short time in Jenney’s office after the great fire of 1871, left him soon after and studied architecture at the École des Beaux-Arts in Paris in the mid 1870s, after which he returned to Chicago and his comprehensive building activity evolved.[80]

[79] Condit, 1964, p.116.

[80] Sullivan, 1956, p.202f. and 213, as well as Condit, 1964, p.30ff.

W.L.B. Jenney: Home Life Insurance Building (1885), prototype of the "Chicago frame". (Condit, 1964, illus. 44)

When the Home Insurance Company appointed Jenney as chief architect in 1884, the 52–year-old architect took on his first important building assignment representing the starting-point of a very successful career. This was the first time he was able to design a very large and high building and moreover apply the newest technology in iron skeleton construction as well as the newly invented Caisson Foundation.

The office building had a cellar and was nine stories high; in 1891 two stories were added. George B. Whitney, who studied civil engineering at the University of Michigan and built his reputation as a bridge engineer for the Mississippi River Commission, collaborated on the project as civil engineer. This was

Two construction drawings by Jenney published in "The Engineering Record" on the Fair Building: load-bearing structural skeleton made of steel sections, brick surface fireproofing, and a connecting typology of supports and girders. (Jenney, Nov. 1891, p.388)

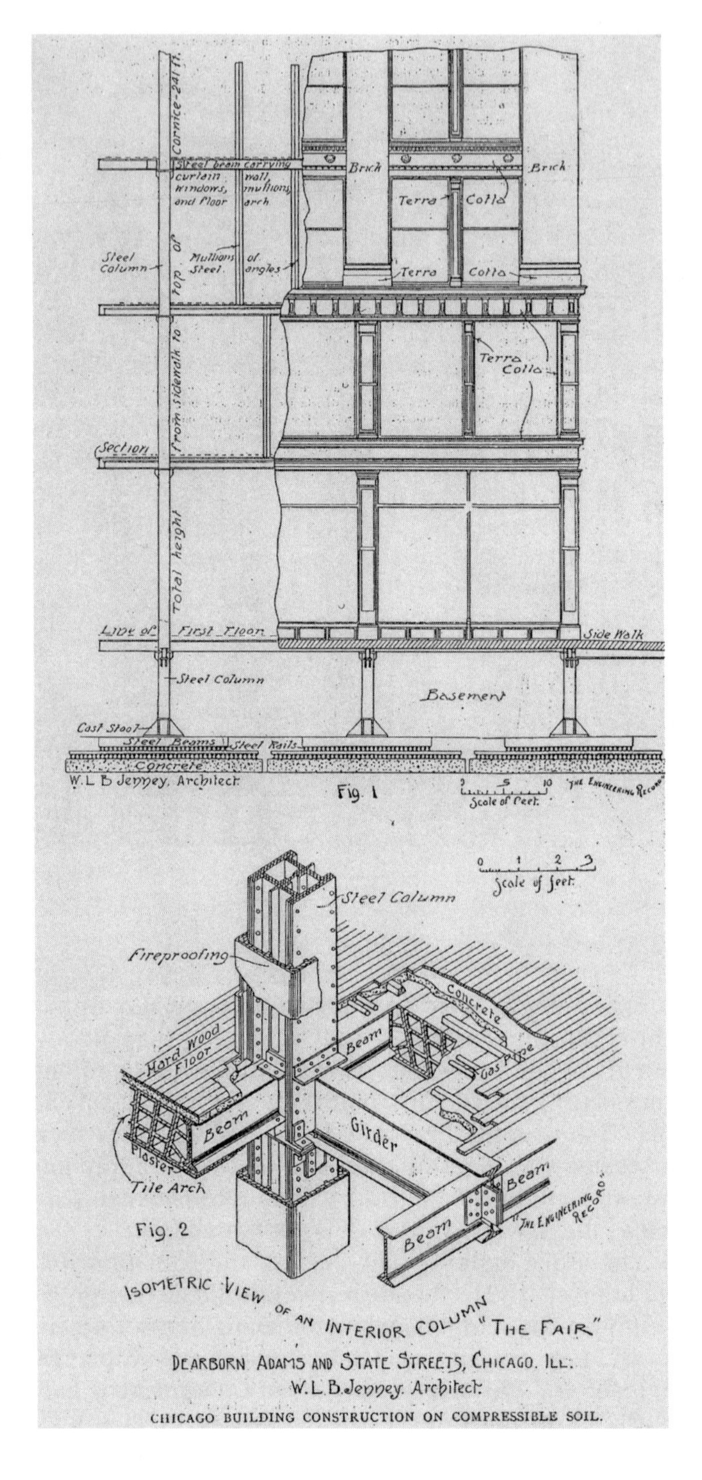

the first time that a skeleton construction entailing a system of supports and girders made of customary steel sections was erected in Chicago; furthermore, the building's jointing solution was Jenney's own invention, which was published in an illustrated article in "The Engineering Record" at the end of 1891, concurrently also in the London edition, using the example of his Fair Building.[81]

[81] Jenney, Nov. 1891.

The Home Insurance Building with its logically consistent application of the steel skeleton was a milestone for many engineers and architects of the time. In his account of this building's construction Frank A. Randall cites the German engineer and graduate of the Stuttgart Polytechnical School, Charles Louis Strobel, who at that time did innovative work in America in iron technology and who emphasized the walls in their function as non-supporting, simple division and in-filling in contrast to the supports, which led to a considerable redesigning of structural parts. D.H. Burnham was also cited, who praised Jenney's inventive spirit, whose equally successful architectural firm did not use steel skeleton construction until 1888, yet further modified it until its technical and aesthetic zenith was reached in the Reliance Building (1894/95).[82]

[82] Randall, 1949, p.105–107.

Jenney was an inventor, architect and propagandist at the same time. On the occasion of the international architectural congress of 1887 held in Brussels, and in 1893 in the context of the World's Columbian Exposition in Chicago or in 1900 in Paris, Jenney held lectures on the use of steel skeleton construction in Chicago, now commonly called the "Chicago frame". In a principle article titled "An Age of Steel and Clay" published in "The Inland Architect and News Record" of 1890,[83] he explained the "Chicago frame" and placed it in a cultural-historical context. In his article Jenney points out that the transition from iron to steel (rolled and sectional steel), thanks to Bessemer processing, would not only cause a railway boom but also cause buildings in Chicago to increase in height; Carnegie, the same company who introduced the Bessemer process in the United States, had already delivered I-girder beams to be used in the Home Insurance Building. In order to become fireproof, however, the steel parts had yet to be surfaced. Although first stone, then bricks

[83] Jenney, 1890.

and finally hand-made terracotta were used for this, in this article Jenney brings his readers' attention to a new method that mechanically produces fired clay or terracotta of a consistent and high material quality, thin-layered and of high density, with standardized forms and a smooth and self-cleaning surface.

Jenney also takes the opportunity to announce the establishment of a new kind of factory that would mass produce these surface elements 24 hours a day, 365 days a year to meet Chicago's demand; for this a production line in the form of a covered channel was developed, in which raw material is fed into the one end and the finished product emerges from the other. This procedure's decisive advantages lie in its continuous, controllable, swift and economic process of production. Similar processes were also to benefit steel production, which around 1890 could not meet the increasing demands (quantity and delivery time) made by architects and engineers for the construction of Chicago's skyscrapers: "And we enter upon a new age – an age of steel and burnt clay."[84]

[84] Jenney, 1890, p.77.

Such statements, which are an expression of Jenney's activity as director of one of the larger architectural firms, of his inventive spirit and his drive to make public what he knows, not only reveal the sentiments of the modern engineer of the era of steel and glass but also reveal the *Ingénieur industriel* consciousness. His school, the École Centrale in Paris, educated and inspired architects, engineers, inventors, scientific researchers, organizers and entrepreneurs simultaneously. Jenney belongs to this new type of universal organizer. He left his mark on modern Chicago, and in the offices that followed – up to Mies van der Rohe and Skidmore Owings & Merrill – he had an unmistakable impact.

In an article published in the "Inland Architect" of 1889 Jenney paid the highest tribute to his French architect colleagues of polytechnical and industrial background and pointed out the perfection of their working method and the didactic value of their presentation technique; in contrast to the "beautiful drawings" of "entirely unconstructible" buildings by École de Beaux-Arts students, the former stand out in Jenney's opinion by exhibiting the best detailed drawings, in grasping the building program, explaining the interrelationships and in the practical success that their firms enjoyed.[85]

[85] Jenney, 1889, p.7–9.

A work equally inspired by Europe is the Horticultural Building constructed for the international Columbus Exhibition in Chicago in 1893 by Jenney and Mundie; in this context it was an example of one of the most outstanding ingenious architectonic achievements. This greenhouse measured 330 meters in length and 82 in width (the Crystal Palace in London measured 616 x 144 meters); a central domed hall, the "crystal dome" measuring 38 meters high, rose above the spacious pavilion grounds, which as a matter of interest was given a historic facade evoking Venice of the Renaissance. In contrast to the circular hall, the glass greenhouse architecture with its sectional steel as corresponding to the standard of the time is only visible in the interior.

Overall grounds of the Horticultural Building as a combination of modern industrial steel/glass architecture and historicizing facade; after a drawing by Jenney and Mundie. (Jenney, June 1891)

Besides being influenced by British and French architectural engineering, Jenney's work also exhibits unmistakable elements of American building tradition. This regards the "balloon frame" in timber-frame construction, which was applied for the first time in an important building by Augustine D. Taylor in St. Mary's Church in 1833, three years after Chicago was founded.[86] The balloon frame replaced the wood construction methods imported from Europe as a genuine American system of construction; it spread

[86] Condit, 1982, p.43 and Cavanagh, 1997, p.5–15.

Jenney and Mundie's large exhibition building of horticulture for the World's Columbian Exposition of 1893 in Chicago; central domed hall from the interior and as unfinished structure; compare this netlike patterning of the dome construction with that of the Halle aux blés in Paris of 1811. (Bancroft, 1894, p.25 and 41)

quickly across the Middle West, allowed settlements to be erected quickly and was used for establishing later cities (such as Oklahoma in 1889), and still today is a common form of construction.

The balloon and platform frame is more than a method, it represents a pragmatic form of thinking and working, which met the needs of rapidly expanding settlement. Didactically organized handbooks facilitate learning and grasping the system quickly and efficiently. The balloon frame, as a "Yankee technology" acted as a model for Jenney's invention of the steel frame system for Chicago's skyscrapers.

Thus did three lines of development, intellectually and pragmatically converging in the person of Jenney, result in the establishment of the "Chicago School": The tradition of iron skeleton construction developed for British textile industry initially separated the outer wall from its load-bearing function and broke up the roof construction into lighter layers; the French tradition of engineering then resulted in perfecting the use of steel in supporting structures and in reapportioning the quantity of material used for the components; the American approach introduced a new principle of economy. Therefore, in Jenney's work the French industrial school of thought embodied by the École Centrale is combined with a fundamental American principle, namely a combination of standardizing construction components (interchangeable parts) with pre-fabricated elements in a secure workshop and mass produced material of a consistent quality coupled with fast and easy assembly; in other words, precision combined with speed – a precursor to the production maxim that began with automobile production, "time is money".

Armand Moisant

Moisant (1838–1906) graduated from the École Centrale as "Constructeur" in 1859 and subsequently worked as an engineer and renowned businessman in metal construction production as well as a agricultural engineer. Like a true "Centralian" after receiving his diploma he also supported his school and other young engineering students as president of the Association amicale des ancien élèves.[87]

[87] Guillet, 1929, p.252 and 280.

In 1866 Moisant founded a workshop for metal construction together with a firm for plumbing facilities, which turned into one of the largest metal construction firms in France within 20 years. Between 1868 and 1889 his company's production of metal structures soared from 2,000 to 10,000 tons a year.[88]

[88] A short treatment of the firm in: Lemoine, 1986, p.303.

In 1887 the company was called Moisant, Laurent, Savey & Cie., in 1897 Henri Garnier, another Centralian, joined the firm; and in 1902 the company became a Société Anonyme des Ateliers Moisant, Laurent, Savey. Moisant's first business partner, Édmond Laurent, also received his diploma from the École Centrale (1879) after first attending the École des Arts et Métiers in Châlons and subsequently acquiring practical experience at Moisant's production sites. He entered the École Centrale at Moisant's behest; after his diploma he returned to the firm.[89]

[89] Guillet, 1929, p.315.

Besides building railway stations (i.e. the Gare de Lyon in Paris 1896–1898), covered markets (Marché de Sens) and bridges (about 150 until 1935, including those of the Compagnie du Chemin de Fer du Midi

Construction halls of the Moisant-Laurent-Savey forge and metal workshops around 1885. Industrial work organization combined with functional architecture. (Lemoine, 1986, p.212f.)

over the Rhone near Lyon), Moisant and his firm erected a few important buildings that were to influence the history of 19th century architecture.

One of these is the prominent Menier chocolate factory in Noisiel-sur-Marne east of Paris (since 1996 the French headquarters of Nestlé). It was constructed around 1871/72. Émile Menier had already begun planning the factory's construction in 1864 and together with his architect, Jules Saulnier, he developed a building complex that was based on functional spatial and supporting structures which corresponded to the production process and its conditions. In 1869 Saulnier decided to erect a steel skeleton structure for the central mill with facades made of crossed metal rod grating similar to those used in bridge building. This was to act as a suspension system for the dynamism and vibrations produced by the modern turbines installed in the factory, similar to the solutions realized in British textile industry. It was the first time that a pure steel structure and a non-supporting building skin was developed in France like the ones

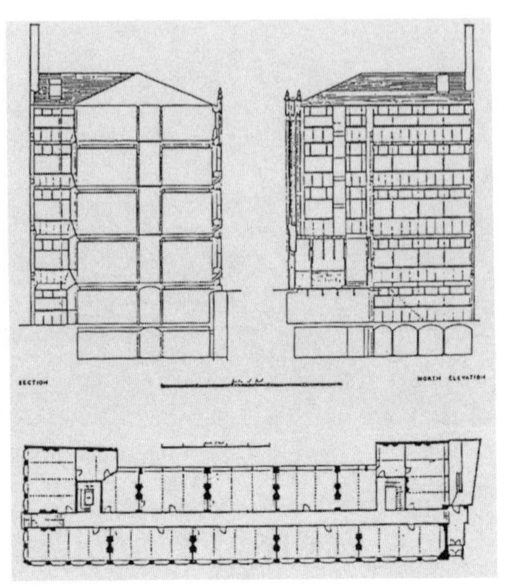

The "Oriel Chambers" office building on Water Street by Peter Ellis jr. in Liverpool (1864). (Hughes, 1969)

used by the architect Peter Ellis in 1865 in Liverpool and as of 1879 by Jenney for Chicago's skyscraper construction.[90] [color plate 8]

Saulnier cooperated with Moisant in order to realize and work out the engineering aspects of the facility. The building is an invention that statically, structurally and technically combines the metal frame structure (support structure) with an attached facade, i.e. a non-supporting facade made of diagonal rods attached to and reinforcing the inner structure on all four outer sides and decoratively adorned with glazed bricks. The entire building rests on a platform that resting on four double pillars and spanning a river – an astonishing early anticipation of the pillared building ("Maison sur pilotis") combined with a steel frame construction ("Ossature à sec") and a free facade ("Pan de fer" or "Pan de verre"), as was later realized for example by Le Corbusier in his student building Pavillon Swiss in Paris (1930–1932) or in his Maison La Clarté (Wanner) in Geneva. [color plate 9]

A more comprehensive application of this construction system was subsequently undertaken by the firm Moisant-Laurent-Savey for the storage house Les Grands Magasins du Bon Marché in cooperation with the architect Charles-Louis Boileau and Gustave Eiffel (1878). Further works include the Halle of the Comp-

[90] Marrey, 1989, p.44.

The Menier chocolate factory in Noisiel-sur-Marne by Jules Saulnier (Architect) with Armand Moisant (Engineer and Metal Construction); Building dates: 1871/72. Support construction, steel skeleton as supporting system and attached facade with diagonal rods: a prominent forerunner of modern architecture. (Lemoine, 1986, p.55)

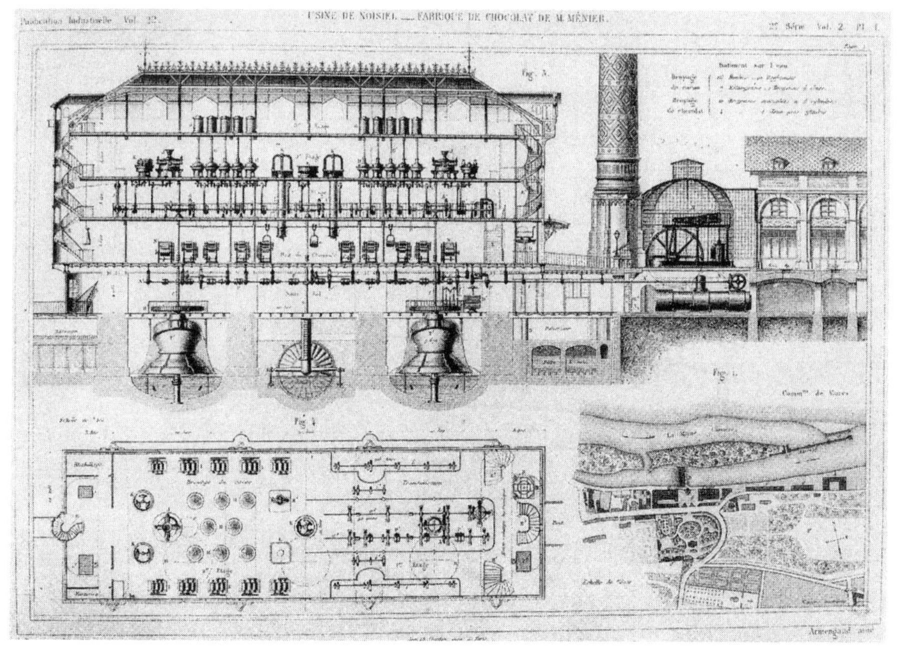

Building structure, construction and production form a unity. (Marrey, 1989, p.45)

toir National d'Escompte (1883–1907), the central domed hall of the International Exposition in Paris of 1889 and the most important section of the Grand Palais des Beaux-Arts, the Palais des Vêtements for the World Fair in Paris in 1900.[91]

As an engineer and in cooperation with an architect, Moisant was therefore capable of solving structural and architectonic problems and as a businessman he was also able to adequately respond to the assignments being formulated. Tight budgeting and the already then short periods allowed for construc-

[91] Lemoine, 1986, p.77 and 236.

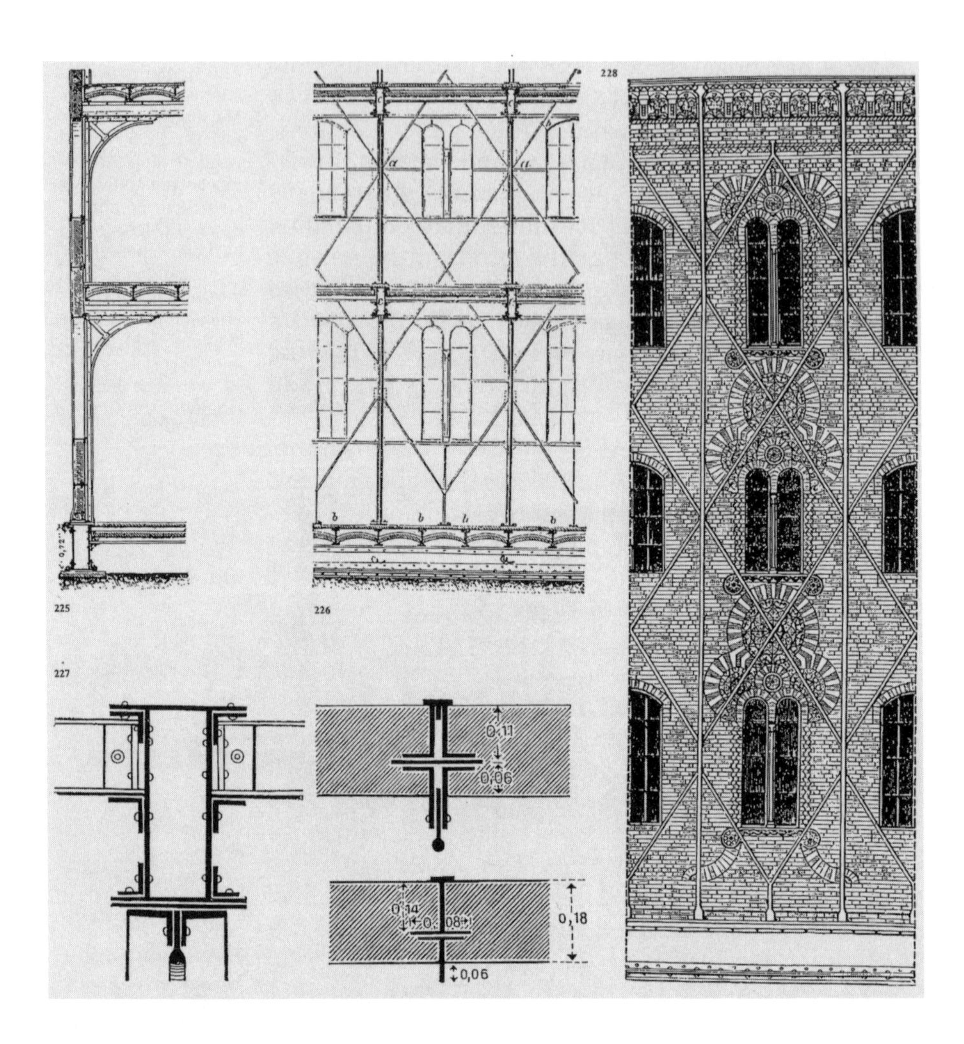

Cross and longitudinal sections of the Menier chocolate factory; the joint of the roof girder and horizontal sections of the outer wall (support and diagonal member) – a combination of "Ossature à sec" and "Pan de fer" as used by Corbusier 60 years later in housing construction. (Roisecco, 1973, p.293; Assembly by: Schild, 1983, p.79f.)

tion compelled the rationalization and standardization of procedures, thereby making production, transport and assembly more efficient; hence, the kind of regular and systematic procedural methods that were being practiced by the Centraux in their education, projects and studies and in their analysis of specific production facilities in field and vacation work.

Victor Contamin

Contamin (1840–1893) is one of the most famous engineers of the 19th century and one of the most renowned graduates to come out of the École Centrale, where he received his diploma in 1860 as a "Mécanicien"; a passenger steam boat represented his final project. He presented his "Bâteau à vapeur" in seven large-formatted plans, with ground plans, sections, elevations, structural projections and detailed drawings as well as precise renderings of the steamboat's motor, of its function and its individual components; here Contamin worked out the program in various scales to provide not only an overall view of the entire ship but also to exhibit details, such as the ship deck's planking or the techniques of assembling individual parts. It was written in quill on precious paper and rendered with an Indian ink wash: a seven-part "image" portraying the interplay between the machine, its operative processes, construction method, material and commercial use as a kind of "Tableau" representing the state of development of engineering technology and industrial aesthetics of that period.[92]

[92] The work is stored in the archives of the École Centrale, which unfortunately only preserved work produced every five years (i.e. 1835, 1840, 1845 and so on).

Considering that the final project was completed within ten weeks and that a detailed and precise description of the facility and construction, including static, mathematical and physical calculations, etc. were to be compiled concurrently (about 30 handwritten pages in Indian ink), and comparing these with final projects completed in a comparable amount of time at schools of architecture today, the Centraux's achievements are astonishing. This quality of work has not only characterized the École Centrale but also other school types of the time, such as the École des Ponts et Chaussées[93] or the École des Beaux-Arts, where exceptional projects were completed, even if in other areas of study.

[93] Cf, student projects treated in Picon, 1989.

Some of Contamin's teachers were Léonce Thomas, successor to Colladon in steam engine construction, Mary for architecture, Perdonnet for railway engineering, Ferry in the field of iron metallurgy and Delacroix for the newly established course on industrial economy as of 1856.

After his studies Contamin first began to work as a simple draftsman in 1863 at the Compagnie du Chemin de fer du Nord, and then slowly began to assume more responsibility and new tasks until he was appointed senior engineer in 1890. In the process of his work Contamin was a busy innovator of, among other things, railway track material and loading or coupling techniques. In this period of time he also produced a scientific treatise on improving static calculations of metal constructions, which was to aid him in his work for the Paris World Fair in 1889.[94]

[94] Guillet, 1929, p. 312.

From 1873 until 1893 Contamin was a professor at his school for applied strength theory and material technology (mechanics department), for which he published his lectures in 1878 titled "Cours de résistance appliquée", which became widely popular. Between 1874 and 1891 he was president of the École Centrale's student aid fund and of the Société des Ingénieurs Civils de France in 1890.[95]

[95] Guillet, 1929, p.116, 236 and 264.

Yet Contamin's most important work was completed for the World Fair of 1889 in Paris. The Polytechnicien Adolphe Alphand (1817–1891), the exhibition's director, appointed Contamin as senior engineer and chief inspector of all metal constructions. Contamin immersed himself in work over a period of three years (this is after all when his fellow student Eiffel created his tower), although becoming internationally famous he ruined his health; he died in 1893 at the age of 53. The most important structure next to the 300–meter tower was Contamin's Palais des Machines.[96]

[96] Guillet, 1929, p.312.

The Palais des Machines was created in collaboration with Contamin as engineer and the architect and Beaux-Arts student Charles-Louis-Ferdinand Dutert (1845–1906). Dutert attended the École des Beaux-Arts in 1863/64 and in 1869 received its coveted Rome Prize.[97] Besides establishing Alphand as director of the International Exposition of 1889, Charles Garnier was appointed as architectonic adviser. Like Dutert, Garnier was also recipient of the École des Beaux-Arts' Prix de Rome (1848) and became internationally famous particularly for his construction of the Paris Opéra. In this respect every single leading school of Paris was prominently represented in this exposition held on the Champs de Mars.

[97] Durant, 1994, p.4–9.

Contamin's feat of engineering in the Palais des Machines entailed the calculation and construction of a completely new form of arcuated support system comprised of three-hinged trussed arches. The arch spanned 115 meters in width and was 45 meters high; it was joined in the middle by a hinge and supported on the ground also by a hinge on each side fixated on a plinth tile. Every hinge bolt carried a vertical load of 412 tons and a horizontal thrust of 115 tons.[98] The trussed arched girders decreased in size downwards, forming pointed ends resting on visible, moveable plinth tiles. In testing the accuracy of Contamin's calculations, Angus Low of the Firm

[98] Marrey, 1989, p.59.

The Palais des Machines at the Paris Exposition of 1889; novel three-hinged trussed girders on moveable plinth tiles span the 115-meter-wide hall; view into the side galeries. (Durant, 1994, p.41 and 40)

Ove Arup Partners in London produced in a modern computer simulation a similar distribution of forces as the one determined by Contamin almost one hundred years ago! Moreover, Low mentions the steel's exceptional material quality and that the riveting parts were produced on site (32,000 rivets per girder), and still un-cooled, were compressed by two simultaneous hammer blows so that they would solidify while cooling and make exceptional and permanent fixings.[99]

[99] Durant, 1994, p.56f.

Giedion mentions in "Building in France" (1995) that because the girders were so light and lacking filling material a new kind of perception was demanded: "The eye of the contemporary onlookers felt insecure and disturbed as the light pouring in from above swallowed up the thin latticework. The vault optically attained an unusual hovering state."[100] No longer were columns used as those built in the Galerie des Machines for the 1878 Expo (constructed by Henri de Dion, who also received his diploma from the École Centrale in 1851), or for the following Expo of 1900 (Galerie des Machines). The 430–meter-long machine hall was formed by 20 three-hinged trussed arches, which were reinforced by an exterior secondary metal structure along the length of the hall [on both sides], serving also to support the entire structure's glass roof. It took six months alone to complete the roof.[101]

[100] Giedion, 1995, p.139.

[101] The commission to build the hall girders was given to two different firms who each worked from one side towards the middle: the Compagnie de Fives-Lille and Cail et Cie.; their assembly techniques were varied; cf. Durant, 1994, p.23f. and the firms' biographies: Lemoine, 1986, p.301f. and 299.

The World Exposition of 1889 in Paris therefore produced two of the advanced industrial age's most outstanding achievements of architecture and engineering, in which graduates of the École Centrale were decisively involved; Eiffel's "Tour de 300 mêtres" and Contamin's Palais des Machines. Just as the Eiffel tower was to be the tallest structure yet built, the Contamin/Dutert team sought to break a horizontal record by spanning an exceptionally large space. The hall was designed as a light and columnless structure in order to reduce the load placed on the inferior building ground.It was to enclose an awe-inspiring space in which the most advanced machinery was to be presented, thus demonstrating the productivity of French metallurgy and steel industry. The exhibits were to be freely accessible to

"Ponts roulants": Moveable viewing bridges for the public in the machine hall of the Paris Exposition of 1889. (Durant, 1994, p.23)

the public and also viewable from above; for this purpose mobile panorama bridges were designed and installed ("Ponts roulants") above the exhibits, vertical vantage-point lifts were also built.[102]

[102] Lemoine, 1986, p.228.

The Engineers of Concrete
Although concrete was already known as building material for some time and was, as already mentioned, treated in Mary's textbooks around the middle of the century as a compression-resistant material for foundations, as tamping-concrete for walls and pillars and for use in weight-carrying structures, the true inventors and pioneers who introduced iron-reinforced concrete in building design and construction as support structures with compression and tension zones above and below respectively were members of the Coignet family. It is no coincidence that the statically beneficial interrelationship between concrete and iron was a discovery of chemistry and that the pioneers of concrete were to be found within the circle of the Centraux.

François Coignet (1814–1888) is considered the pioneer of "Béton aggloméré". He managed a construction and chemical firm. Already in 1855 he received a patent in Great Britain for cross-jointed iron rods to fortify concrete girders. In 1861 his treatise

"Béton aggloméré à l'art de construire" was published. It is on this basis that concrete construction became industrialized. On the occasion of the Paris International Exposition of 1867 Coignet first designed a concrete structure to cover-over the subterranean service rooms built in front of the pavilion buildings.[103]

[103] Giedion, 1995, p.150, and Fig.39, p.129.

While François Coignet was doing his pioneering work, *Édmond Coignet* (1856–1915), his son and graduate of the École Centrale, led the way in the industrial application of reinforced concrete. Beginning in 1888 he endeavored to improve concrete technology by perfecting concrete production, or sintering methods ("Béton aggloméré"), and extending the industrial application of this novel material. In 1892 he constructed a subterranean elliptical concrete gallery near Achères that was over 5 meters wide and 2.4 kilometers long.

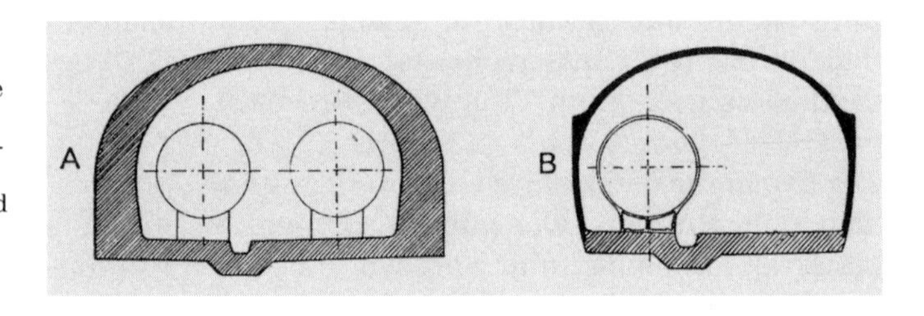

The comparison between a traditional masoned conduit-type sewer and an enclosing wall made of reinforced concrete; after a proposal by Édmond Coignet (1892). (Singer et al., 1970, Vol.V, p.490)

Together with Napoléon de Tédesco of the "Société des Ingénieurs Civils de France" Coignet submitted a report in 1894 on the method of calculating iron-reinforced concrete, which is still widely practiced today.[104] Coignet is one of the innovators of the concrete column and the earthquake-proof concrete structure, which he was able to install in many countries. Besides concrete construction he also was engaged as an engineer in waterworks, road and railway construction, electrical installations and more. Coignet also developed the first lighted fountains with waterworks for the Paris World Exposition of 1889.[105] Urged by Édmond, the Coignet family established a national concrete prize, "Le Prix François Coignet".[106] Coignet, in his function as vice-president of the School Council, was also able to support his school.

[104] Straub, 1992, p.271.

[105] Guillet, 1929, p.310f.

[106] Guillet, 1929, p.264.

With the Firm Hennebique a second important "school of concrete construction" became established in France. *François Hennebique* (1842–1921) invented the first iron-fortified "Tee beam" (1892)[107] and founded one of the most internationally famous businesses dealing in steel-reinforced concrete with branches and licensed partners in many countries. In 1914, for instance, the company maintained 64 branches and 479 licensed partners in France and further 246 in 38 countries. The "Bétons Armés Hennebique" firm existed until 1967 and could look back on an Oeuvre of approximately 150,000 projects.[108] The Hennebique building system (Système Hennebique) entails a monolithic employment of supports, girders and roof slabs joined by a continuous system of iron reinforcement; Hennebique's formal and characteristic joint design as an inclined joist further acts to stabilize the entire structure. Moreover, this "skeleton construction method" also guarantees fireproofing and allows for spacious functional and flexible utility without great interference from structural elements.

[107] Straub, 1992, p.272.

[108] On the history, importance and activity of the Hennebique firm, cf. Delhumeau, Gubler et al., 1993; here p.186.

One of the first modern buildings to be erected according to the Hennebique building system is the residential building on the Rue Franklin in Paris by Auguste and Gustave Perret (1902/03). Giedion describes it as the "first undisguised skeleton-frame

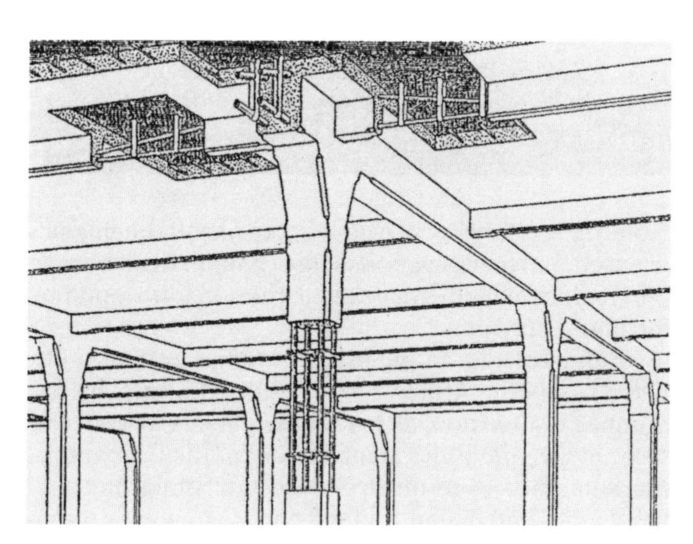

"Système Hennebique" of 1892; the re-arrangement of material quantity (combination of supports, girders and thin floor/roof slabs; continuous reinforcement) while also increasing efficiency (load-bearing capacity, fire-proofing). (Delhumeau, Gubler et al., 1993, p.46)

1

2

Fiat Lingotto in Torino (Engineer: Giacomo Mattè Trucco; 1916–1920); realization of the support structure according to the "Système Hennebique" by the licensed firm of Porcheddu: the company and the structural system guaranteed a speedy and inexpensive completion of Italy's largest construction site at that time; the industrial principle of construction "time is money" corresponded to the American model picked up by Giovanni Agnelli from his meetings with Henry Ford (around 1910) in Detroit; the Ford Old Shop in Highland Park, Detroit (Mich.) by Albert Kahn (1909) probably stood as model. Illus.1: Production floor, Illus. 2–3: South ramp. (Contemporary photographs from: Olmo, 1994, Fig.72, 5 and 7)

building in ferro-concrete."[109] Among Hennebique's licensed partners was also the Italian firm Società Porcheddu, which constructed the Lingotto building for Fiat in Torino.[110]

Hennebique also published a company Journal titled "Le Béton Armé" as of 1898. He employed a number of Centraux as leading engineers of concrete technology and in the management of the firm and its markets; they were involved in establishing the company and contributed to the firm's competency, inno-

[109] Giedion, 1995, Fig.68, p.154; cf. also the periodical "Rassegna" No. 28, 1986, esp. p.32ff.

[110] Olmo, 1994, p.58–62; with reference to America, cf. Banham, 1986, p.236–253.

3

vative activity and reputation. The first of these was *Maurice Dumas* who graduated from the École Centrale with a diploma in 1892 and who calculated, designed and built one of the first iron-reinforced concrete bridges (near Châtellerault). Guillet also mentions a number of other engineers who received their specialized diploma in concrete from this school between 1895 and 1897 and who were employed by Hennebique.[111] One area in which Hennebique's engineers also specialized in was in earth-quake proofing.

[111] Guillet, 1929, p.311.

Another graduate of the class of 1897 from the École Centrale was *J.-T. Gueritte*, who moved to Britain and was director of J. Mouchel & Partners, one of the major British pioneers of iron-reinforced con-

Earthquake-proof construction with Hennebique; innovations, great feats of engineering and extensive regional presence through licensed partners as a characteristic profile and principle of promotion: The "Système Hennebique" was more than a method of construction. (Delhumeau, Gubler et al., 1993, p.79)

crete.[112] In the same year the Mouchel firm established a branch in London in 1897, Guéritte introduced the Hennebique building system into Britain. Mouchel was the only concrete construction company in London until in 1904 a student colleague of Guéritte, Coignet, also opened a branch there.[113]

Many developments and innovations in France (involving also the Polytechnicien Eugène Freyssinet, diploma in 1899) as well as in America and Britain (Hyatt in New York, Wilkinson in Newcastle upon Tyne and others)[114] improved iron-reinforced concrete as a material, supporting structure and method of construction, "but" – writes Giedion – "the decisive step that would enable a new means of architectural design to arise from an ancillary material, from a construction detail,

[112] Guillet, 1929, p.412 and 280.

[113] Singer et al., 1970, p.490.

[114] Elliot, 1994, p.166ff.

was taken by Fr. Hennebique."[115] The first international manifestation of concrete took place in the Paris Exposition of 1900, where the Coignet and Hennebique firms were represented in particular.[116] The century of concrete thus opened on time and replaced the previous one in which iron dominated architecture. Here too, graduates of the École Centrale played a leading role.

[115] Giedion, 1995, p.150.

[116] Elliot, 1994, p.182.

Engineers of Services and of the Technical Equipment of Buildings

Courses on modern heating installation is a further didactic innovation of the École Centrale. The foundations were laid by *Eugène Péclet*, one of the school's founders and first professor of industrial physics. Like Perdonnet, Colladon or Mary he is a member of the pioneering generation of the Sciences industrielles and of the original École Centrale school of thought. Péclet completed his studies at the École Normale Supérieur, was recommended by his teacher Ampère in 1815 to be appointed to the chair for physics in Marseille, where he established France's first municipal courses for engineers, technicians and foremen from factories and manufacturing companies. As a liberal spirit he fell from grace during the Restoration Regime, in 1827 he lost his teaching assignment and subsequently moved to Paris. Under Minister Martignac, who supported the project to establish the École Centrale, Péclet was once again appointed a chair at his former school. His teaching activity lasted from the year the school was established in 1829 until 1857, at the time the school was transferred to government control.

His successor ad interim (1857–1865) was *Léonce Thomas*, who graduated with a diploma from the École Centrale and who worked as a Répétiteur under his teacher Péclet. However, as of 1835 Thomas had already been teaching the science of steam engineering as successor to Colladon (until 1869).[117] His innovations include an early form of the steam boiler (Chaudière mixte; in collaboration with his fellow student Camille Laurens, who received his diploma a year earlier); the heating and ventilation system in the Lariboisière Hospital after 1852 represents his most important practical work.[118] The hospital had just been completed when the decision was made to

[117] Guillet, 1929, p.138.

[118] Guillet, 1929, p.317.

install a modern system of heating and ventilation. The hospital administration then appointed Arthur Morin as an expert for this project. Morin was a Polytechnicien (diploma in 1813), a military engineer and general; between 1849 and 1880 he directed the Conservatoire National des Arts et Métiers and taught and experimented at this unique school.

The heating and ventilation system was to improve the hospital's hygienic conditions and indoor climate. It was a time of heavy epidemics, of cholera for instance, which struck Paris around 1832 and also affected the École Centrale. Lariboisière was to be a bulwark against the cholera epidemic. Yet the volumetry of the building's different wings and pavilions, and the dimensions and complexity of the entire hospital facility posed the most difficult and comprehensive problem that had to be mastered at that time with respect to technical building equipment. Morin preferred to use a ventilation system based on suction and hot-water heating similar to the ones installed at the Conservatoire in 1849 (facility according to Léon Duvoir-Leblanc); however, he wanted to combine this with a proposal submitted by the three Centraux engineers Grouvelle, Thomas and Laurens, which provided for a pressurized ventilation and a steam-powered heating system. These different systems were each installed into one of the building wings in order to distribute the risks and be able to compare their efficiency.

The systems installed therefore not only fulfilled a certain function but also represented a scientific experiment to test different theories. Although the Labriboisière model did not prove itself successful in the cholera epidemic immediately following in 1865, it did so not because of faulty scientific theory or of a lack of ability to function as a system but because it was negligibly handled by the hospital's technical personnel, which partly boycotted the system's control instruments. This novel technology hat yet to master the prevailing consciousness! The instruments were of greatest importance because little was known then about the movement of air in heated rooms.[119]

[119] Le Moël, 1994, p.178.

Morin also considered the Conservatoire to be an arena for scientific research and for testing new systems and apparatus. The first system that was in-

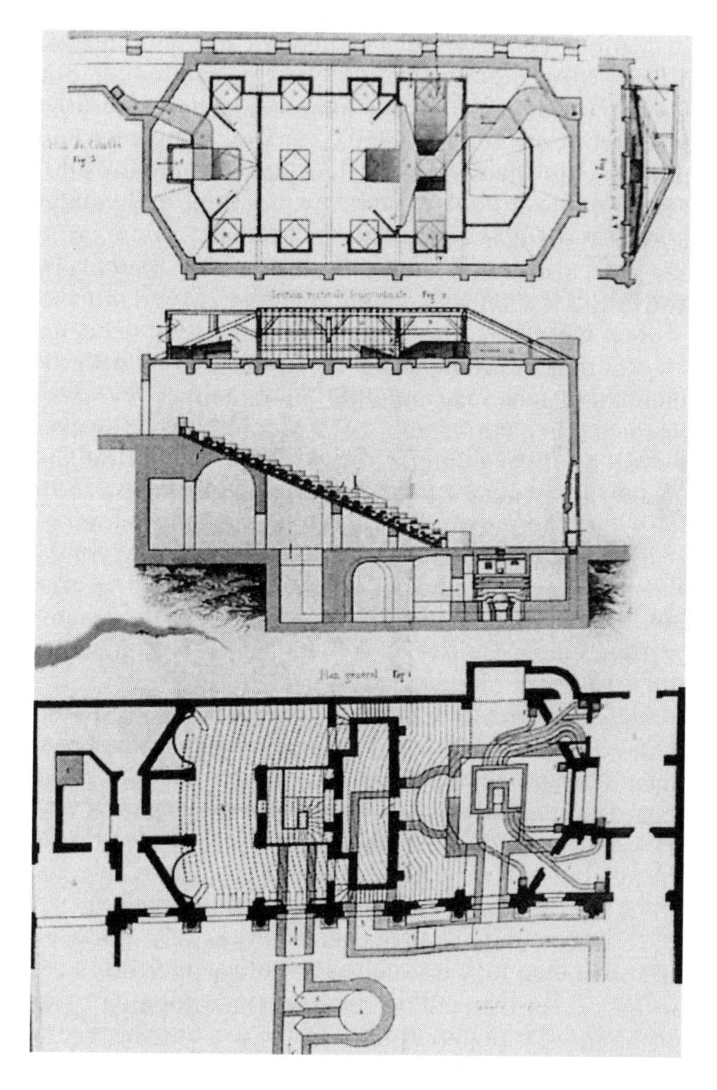

An example of early domestic service equipment systems as developed by pioneers of the École Centrale: Heating and ventilation system (roof-truss, longitudinal section and cellar) of the Conservatoire Nationale des Arts et Métiers' large auditorium, object of experimentation by Arthur Morin (State: 1865). (Le Moël, 1994, p.174)

stalled between 1849 and 1851 had to be modified in 1853 and completely refurbished between 1861 and 1863. Morin subsequently continued to replace individual components in order to check the function of systems and their parts. The starting point for this interest in improving the comfort of domestic installations and technical equipment ("Économie domestique") and for the development of this technology was the pavilion dedicated to this theme built for the International Exposition of 1855 in Paris.[120]

[120] Le Moël, 1994, p.180.

Léonce Thomas was also actively engaged in supporting his school and the engineering profession, not only as a teacher and practicing engineer but also as member of the School Council (1856–1862) as well as president of the former student Association amicale. Furthermore, he belonged to the founding circle of the Société des Ingénieurs Civils de France. His son, Max Thomas, also studied at the École Centrale (diploma in 1867) and worked as Répétiteur (1869/70) and interim professor for his father's course in steam engineering.

Another graduate of the École Centrale who was a pioneer of technical building equipment is *D'Hamélincourt*. He completed his diploma in 1841, was also a student of Péclet and inventor of one of the first circulation systems for hot water, which was installed in both the École Polytechnique's technical equipment system and in the administrative wing of the Gare du Nord in Paris, where they remained in service for over 75 years. Moreover, together with Morin, D'Hamélincourt undertook experiments to analyze the movement of air in heated rooms.[121]

[121] Guillet, 1929, p.317.

Louis Ser is also a further Centraux who graduated with a diploma in 1853 and went on to become a pioneer of technical building equipment. Ser took over from Léonce Thomas' interim professorship of industrial physics ("Physique industrielle" in 1865 and was Péclet's actual successor. He kept this position until 1888.[122] His textbook, "Traité de Physique industrielle" was a comprehensive overall review of the state of knowledge and development concerning this core sector of industrialization. He experimented with problems of transmission technology in heating, ventilation, water supply and drainage. His practical work focused in the public service sector where he worked as a leading engineer. As a pioneer of technical building equipment he was also consultant for the firm Geneste & Herscher in their installation projects, for instance for the École Centrale's heating system.[123]

[122] Guillet, 1929, p.138.

[123] Guillet, 1929, p.317ff.

One of the most well-known technical installation firms was the Établissement Grouvelle & Arquembourg. *Jules Grouvelle* from the École Centrale's graduating year of 1861 and Ser's successor as professor of industrial physics (1888–1920) managed this company as of 1861 together with his father, one of the most important inventors of the steam heating system. When his father died in 1865 he managed the business alone until in 1880 he took on the Centralian engineer *Henri Arquembourg* (diploma in 1879)as an associate. Other Centraux joined the firm later, such as Léon Joret who graduated from the École Centrale (1897), Louis Arquembourg (1878) and others. This company brought forth many trailblazing innovations and patents (low-pressure heating, automatic air-pressure regulation around 1890, fully automatic central heating over several stories in 1894, etc.); it became the leading company for technical building equipment and it established branches in various countries.[124] Geneste & Herscher was a further important firm in this sector. The engineers Ser and Fourchotte (diploma in 1870 at the École Centrale) and Étienne Herscher himself were consultants of this firm. Herscher, who developed one of the first low-pressure heating systems to be powered by steam, was also a Centraux (1886).[125]

[124] Guillet, 1929, p.318.

[125] Guillet, 1929, p.319.

Civil Public Construction

In the area of civil public construction (architecture and city planning) other Centraux than those already mentioned should be pointed out, among them *Henri de Dion* (diploma in 1851), metal construction engineer and builder of the 1878 Paris Exposition's machine hall, *M. Denfer* (1861), also graduate of the École des Beaux-Arts and teacher of architecture at the school (1889–1909) after Muller, M. Bourdais (1857), administrative architect of Paris, and M. Demimuid (1858), inspector of the public municipal services of Paris; further, the Swiss graduates and students of Mary, Grenier (1858), director of public works of the city of Lausanne, de Pury (1843), chief engineer for the department of road and bridge construction of Neuchâtel, de Muralt (1851) in the same function for the city of Berne, Wurth (1854) for Geneva and de Montenach (1856) for Fribourg.[126]

[126] Guillet, 1929, p.306–311.

In the literature on architecture and art history in most cases only the architects of such work are mentioned, less the engineers[127] and hardly in any case the firms involved. However, as a result, neither the École Centrale's importance as a school of thought nor its influence in history can be understood and examined.

[127] An exception is Giedion's "Building in France" and the sources mentioned here.

Placing the École Centrale in the Context of the Era

The Educational Model and Technological History in the Historical Context

The industrial model of education invented and developed by the École Centrale on the one hand stands in the polytechnical line of tradition and is to be seen as a continuance of the Modèle polytechnique and the specific principles of the Enlightenment, e.g. unlimited access to education, underlying that model. On the other hand the political shift that occurred after 1830 effected a swift and widely tangible development of industrial forces and of political conveyors of modern economic life in France. The social utopian circles that existed during the Restoration Regime, who in some respect continued the tradition of the "Lobby scientifique" of the revolutionary era and who were equally a dedicated and endangered group, now attained official status as political dignitaries, modern industrialists, respected engineers, etc. Moreover, it is no coincidence that with the subsequent political shift brought about by the 1848 Revolution and the establishment of the Third Republic an honorable society of engineers was called to life: the "Société des Ingénieurs Civils de France".

Between 1830 and 1848 French industry grew to as yet unheard of proportions. The École Centrale was established during this time. Its founders, professors and moving forces were engineers and architects on the cutting-edge of development. From this the interrelationship between the educational model and technological development was formed. A comparable relationship can be observed in the preparations for founding the École Polytechnique and in the school's subsequent establishment phase.

The industrial pioneers who taught at the École Centrale were not only capable of including in their teaching their knowledge and experience of praxis but they also considered the school to be the starting point for their field of research. As a result, the practice courses and application exercises at the École Centrale were of equal importance as instruction in the theoretical principles; in practice work "leading issues" were treated, which becomes apparent in the

Palais des Machines at the Paris Expo of 1889; unfinished structure. (Lemoine, 1986, p.230f)

final projects completed by the school's students. Concurrently, the "didactic challenge" that the engineers and architects as teaching staff were faced with resulted in raising scientific levels and systematic working methods. As an example for this interrelationship the "Polonceau Truss" should be mentioned here again, or Victor Contamin, whose Palais des Machines for the Paris Exposition of 1889 was built during his teaching activity at the École Centrale.

Professionalism and Self-Concept of the Centraux

In the period of industrialization after 1830 and as a consequence of the new educational model established at the École Centrale a whole number of nascent professional fields and branches developed. The type of building programs brought about by the altered historical circumstances (such as railway engineering, mechanical industry, hospital construction, heated and technically equipped urban housing, etc.) and the specific challenges faced by modern architects and engineers (e.g. factories for complex industrial processes, mastering building programs of as yet unimaginable magnitude), as well as the economic conditions of approaching modernity, such as speed, rationality (i.e. in exhibition buildings and the organization of major

TABLEAU COMPARATIF DES COURS ET CONFÉRENCES

TITRE DES ENSEIGNEMENTS	NOMBRE DES LEÇONS						
	1830	1850	1870	1890	1910	1929	
						Génér.	spéc.
Analyse mathématique	70	60	60	33	28	30	—
Mécanique générale				50	43	40	—
Cinématique	—	20	24	—	45	40	—
Géométrie descriptive	70	60	60	49			—
Physique générale	70	60	60	60	54	50	—
Thermodynamique	—	—	—	—	12	40	—
Mécanique des fluides	—	—	—	—	—		—
Physique industrielle	70	60	45	38	38	36	—
Chimie générale	70	60	60	60	55	—	—
— analytique	—	36	60	48	38	—	—
— générale minérale et analyse minérale	—	—	—	—	—	45	—
— générale organ. et analyse organique	—	—	—	—	—	30	5
Minéralogie et Géologie	36	—	30	30	40	18	10
Géométrie descriptive appliquée	70	—	—	—	—	—	—
Théorie des machines	60	—	—	—	—	—	—
Mécanique appliquée (2e année)	—	60	60	55	54	54	—
— — (3e —)	—	60	60	45	41	—	—
Résistance appliquée	—	—	24	22	—	—	—
Constructions métalliques	—	—	—	—	26	34	10 (1)
Machines à vapeur	22	30	35	34	41	45	—
Eléments de machines	—	—	20	21	33	—	—
Construction des machines	30	60	60	51	37	32	3
— — (3e année)	30	60	55	43	40	27	5
Electricité industrielle	—	—	—	30	43	70	5
Constructions navales	—	—	25	—	—	—	—
Eléments d'architecture	—	—	10	30	30	—	—
Constructions civiles	36	—	60	50	45	—	—
— — (1re partie)	—	—	—	—	—	20	—
— — (2e —)	—	—	—	—	—	30	10
Construction et travaux publics	—	120	—	—	—	—	—
Travaux publics	36	—	60	53	51	29	8
Chemins de fer	—	35	40	41	49	—	—
— (matériel et traction)	—	—	—	—	—	17	6
— (voie et exploitation)	—	—	—	—	—	27	9
Exploitation des mines	36	65	40	40	40	31	11
Métallurgie générale	—	—	—	—	15	32	8
— des métaux autres que le fer	—	—	—	—	—		
— du fer	70	65	80	56	38	26	12
Technologie chimique	—	8	30	43	54	—	—
Indust. des silicates et gazéific. des combust.	—	—	—	—	—	26	6
Chimie analytique et industrielle	35	—	—	—	—	—	—
— industrielle	35	80	60	54	—	—	—
Application industrielle de la chimie minérale	—	—	—	—	26	18	7
Applicat. industrielles de la chimie organique	—	—	—	—	25	19	9
Matières textiles	—	20	—	—	—	—	—
Physiologie et histoire naturelle	—	30	20	—	—	—	—
Histoire naturelle industrielle	36	—	—	—	—	—	—
— — et hygiène	—	—	—	25	—	—	—
Hygiène industrielle	18	—	—	—	25	15	—
Législation et économie industrielle	—	—	25	25	25	25	—
Statistique industrielle	18	—	—	—	—	—	—
Conférences sur l'organisation des usines	—	—	—	—	—	15	—
— sur les Essais de matériaux	—	—	—	—	6	6	—
— sur l'Industrie automobile	—	—	—	—	—	9	—
— sur l'Aéronautique	—	—	—	—	—	6	—
— sur les Industries textiles	—	—	—	—	—	10	5
— de Mesures électriques	—	—	—	—	14	—	—
— de Constructions civiles	—	—	—	—	—	5	—
— de Travaux publics et de Topographie	—	—	—	—	—	20	—
— sur la Papeterie	—	—	—	—	—	5	—
— de Comptabilité industrielle	—	—	—	—	10	10	—
— d'Hygiène médicale	—	—	—	—	—	15	—

(1) Constructions navales.

construction sites), compelled architects and engineers to organize their activity more professionally. As a consequence, training and practice became more specialized and tasks were concurrently divided, leading to collaborations between different disciplines.

opposite page: Synapse of lectures and courses offered by the École Centrale for the first hundred years. (Guillet, 1929, p.142)

In the model of instruction developed by the École Centrale students were to receive specialized instruction only in the second half of their studies; however they also continued to attend general and obligatory courses throughout, such as physics or chemistry, which differed from the lower course by their further industrial dimension ("Physique industrielle", "Chimie industrielle", "Construction industrielle", etc.). Never-

A studio and laboratory situation in the École Centrale's model of instruction. (Guillet, 1929, Pl.29)

theless, in time the school developed a greater tendency towards specialization, which has been documented in great detail by the chronicler Guillet in his history of the École Centrale's individual departments.[128] Thus, we are able to determine from the table illustrated here that the number of subjects increased from initially 20 to 50!

[128] Guillet, 1929, p.106–145.

The Centraux understood themselves to be a group of dedicated engineers and architects working at the cutting-edge of social progress, applying themselves to improving the general welfare and comfort and participating in a modern, industrially based nation in the form of an "industrial army". The Ancien Regime's "Corps royal" was replaced by a civil, democratic and republican "Corps industriel". It established societies, published specialized journals and carried this "industrial spirit" out into the world. Wherever Centraux worked they founded national associations and fellowships, helped each other and maintained contact to their points of departure.

"Société des Ingénieurs Civils de France"

Different institutions expressed the kind of attachment that the (private) school maintained towards the state. Since 1833, therefore shortly after the École Centrale was opened, the newly established "Société d'Encouragement pour l'Industrie nationale" provided industrial backing and endowed the school with various stipends. In the revolutionary year of 1848 the "Société Centrale des Ingénieurs Civils" was finally established after being discussed since the second year after the school was founded. First termed "Centrale" in order to place it in proximity to the founding initiative it was quickly renamed into the French Society of Civil Engineers (Société des Ingénieurs Civils de France).

The Société des Ingénieurs Civils de France established in 1848 subsequently became the most prominent professional association of the industrial age in France, and as a reaction to the February Revolution of 1848 it was given the task to secure the profession and its ideals. Even those Centraux with diplomas as "Constructeurs" or "Architect-Engineers" became members of this society. The founding meeting took place on 4 March 1848 in the rooms of the École Centrale. Among the 18 board members 13 of them were Centraux, a predominance that was to last for some time. This society was given government accreditation in 1862 for the diploma title "Ingénieur des Arts et Manufactures". From the very beginning value was placed on including non-Centraux engineers in leading positions, such as Adolphe and Eugène Flachat; the latter was appointed by the founding assembly as the first president of the society of engineers.[129]

[129] Guillet, 1929, p.261–265, as well as Jacomy, 1994, p.351ff.

At length, a few teachers and former students of the school took the initiative in 1867 to form an "Association des anciens élèves" after attempts to establish an "Association amicale" failed in 1833 and 1840. The alumni association served to establish solidarity between the students and graduates, to provide young architects and engineers with a network of contacts and help in establishing their careers and also to provide the school with support from outside; social gatherings in the form of balls and other events were also important. All of this ultimately gave expression to a kind of "esprit de corps" exhibited by modern industrial engineers and architects in a way that is similar to the one developed by the Polytechniciens.[130]

[130] Guillet, 1929, p.13, 14 and 17.

The Extensive Activities of the Centraux
A number of leading scientific and technical *Journals* were managed by Centraux, for instance "La Semaine des Constructeurs", a architectural journal founded by Paul Planat (diploma in 1862) in the year 1876; Planat also edited the periodicals "Vie Parisienne", "Construction Moderne" (as of 1885) and "Encyclopédie de l'Architecture" as well as "Art de Bâtir" and more. Furthermore, the Centraux' broad activity also included specialized journals belonging to the various branches of engineering ("Génie Civil", "La Revue Générale des Chemins de fer", etc.).

With respect to *Education* the Centraux were equally active and well represented, even internationally. Guillet points out that in many countries former students established technical institutions of higher learning. In this respect the school of engineering at the University of Lausanne was founded in 1853 by Jules Marguet (diploma 1840) and Louis Rivier (1843), the first one of its kind in Switzerland even before the Zurich Polytechnical School; they immediately engaged five other student colleagues in order to further establish the school and offer a complete curriculum.[131] In Belgium, Spain, Portugal, Italy, Rumania, even in South America and North America engineers trained in Paris were engaged as teachers, school directors or founders of technical colleges. Jenney's teaching activity in Chicago has already been mentioned. Graduates of the École Centrale were also to be found working in Cuba, Egypt, Australia and so on. Even Chinese schools of engineering such as the imperial school of shipbuilding Fou-Chéou were established and managed by Centraliens.[132]

[131] Guillet, 1929, p.283.

[132] Guillet, 1929, p.281–284.

The Centraux' presence is therefore to be found in all industrial sectors. In addition to the portrayal given above with respect to construction and railway engineering, the three original major subjects of the school's instruction program will be mentioned, namely mechanical engineering, metallurgy and chemistry.

Thus, in the field of *mechanical industry* a number of Centraux were active in major industrial factories, for instance R.P. Mathieu (diploma 1834) who managed the largest French industrial firm Schneider & Cie. in Creusot; M. Avisse (1869) who was general director of Fives-Lille; M. Bougault (1869) of the firm Cail (both businesses participated in the construction of the Palais des Machines in the Paris Exposition of 1889); Paul Ziegler (1896) of the Swiss Sulzer company, etc. Others founded their own businesses such as Moisant and Eiffel mentioned above, or Charles Weyer and Émile Richemond (both 1857). The Swiss Henri-Édouard Dubied (1843) established a factory for mechanical constructions in Couvet in the Canton of Neuchâtel in 1866, which 50 years later employed 3,000 workers; Dubied addressed issues of work organization, especially in the clock- and watch-making industry. World famous Établissements also bore the names of Ingénieurs centraux such as Th. Schlumberger (1861), M. Wehrlin (1878); or the Centraux opened foreign branches in France, such as Édmond Thomine (1884), Étienne Besson (1887) and many more who opened the Société Française des Constructions Babcock & Wilcox.

In *metallurgy* Alexandre Gouvy was the first to introduce the "muck process" in French iron technology – he graduated from the school in 1842. In *chemistry* Daguin also initiated the Solvay process for manufacturing soda ash (he received his diploma from the École Centrale in 1839). The French *textile industry* was also influenced by the Centraux: Dollfus, Lederlin, Boeringer, Schlumberger, Koechlin, Ziegler, Hartmann, etc.. Moreover, many graduates were pioneers of *agricultural science and technology,* for instance Armand Moisant (1859) already mentioned or Arthur Liébaut (1864), who together with Georges (1867) and Grzybouski (1869) introduced and disseminated the steam-powered tractor in France. The activity of graduates abroad has been documented by Guillet in a detailed chapter and in a list that includes the year 1929.[133]

[133] Guillet, 1929, p. 406–415.

What is striking is that during the second industrial revolution graduates of the École Centrale were less extensively active as architects and engineers than first as managers of growing businesses and then increasingly as organizers of productive or administrative societies, thus defining a nascent field of activity.

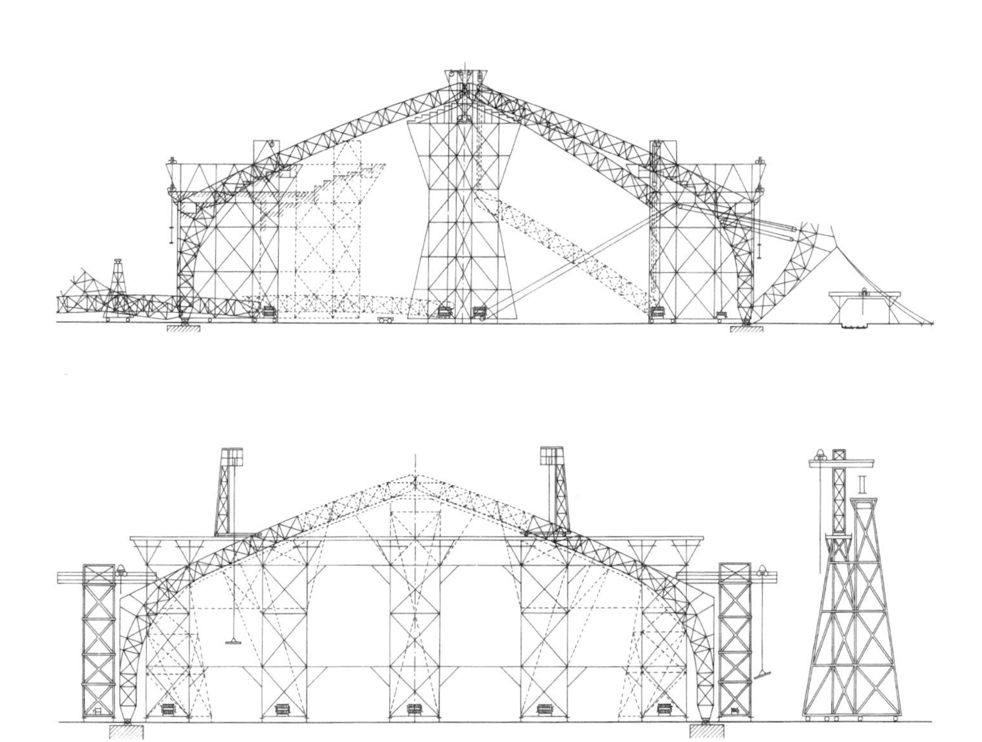

Palais des Machines, Paris Exposition of 1889: Assembly competition between the firms Fives-Lille and Cail. (Durant, 1994, Fig. 75 and 77, p.42–43)

Sources and References to Further Reading

Bancroft, H.H., The Book of the Fair. An Historical and Descriptive Presentation of the World's Science, Art, and Industry, as Viewed through the Columbian Exposition at Chicago in 1893, Vol. 1, New York 1894.

Banham, R., A Concrete Atlantis. U.S. Industrial Building and European Modern Architecture 1900–1925, The MIT Press, Cambridge, Mass., London 1986.

Bissegger, P., Étudiants Suisses à l'École Polytechnique de Paris (1798–1850), in: 'Schweiz. Zeitschrift für Geschichte', Vol. 39, N° 1, Basel 1989.

Boussel, P. (Ed.), Victor Baltard. Projets inédits pour les Halles centrales, Bibliothèque historique de la Ville de Paris, Exhib. cat., Paris (no year).

Brown, D.J., Bridges – Three Thousand Years of Defying Nature, London 1993.

Von Büren, Ch., Der 300–Meter-Turm, Dietikon/Zurich 1988 (French Edition, Lausanne 1988).

Cavanagh, T., Balloon Houses: The Original Aspects of Conventional Wood-Frame-Construction Re-examined, in: Journal of Architectural Education, Vol. 51, N° 1, September 1997, Washington, DC, 1997.

Callot, J.P., Histoire de l'École Polytechnique, Paris 1982.

Colladon, D., Les Travaux Mécaniques pour le Percement du Tunnel du Gottard (Ed. Société Hélvétique des Sciences Naturelles), Genève 1875.

Comberousse, Ch. de, Histoire de l'École Centrale des Arts et Manufactures, Paris 1879 [École Centrale des Arts et Manufactures: archive and library, Paris].

Condit, C.W., The Chicago School of Architecture. A History of Commercial and Public Buildings in the Chicago Area 1875–1925, Chicago/London 1964 (first edition 1952).

Condit, C.W., American Building. Materials and Techniques from the Beginning of the Colonial Settlements to the Present, Chicago/London 1982 (first edition 1968).

Delhumeau, G., J. Gubler, R. Legault, C. Simonnet (Ed.), Le béton en représentation. Le Mémoire photographique de l'entreprise Hennebique 1890–1930, Paris 1993.

Dubin, Ch., Chronique de l'École Centrale, Paris 1981.

Dunham, A.L., The Industrial Revolution in France 1815–1848, New York 1955.

Durant, S., Palais des Machines, London 1994.

Eck, Ch., Traité de construction en poteries et fer à l'usage des bâtiments civils, industriels et militaires, Paris 1841.

École Centrale. Origines et destinées de l'École Centrale des Arts et Manufactures, Paris 1981.

Elliott, C.D., Technics and Architecture. The Development of Materials and Systems for Buildings, Cambridge, Mass., London 1994 (first edition 1992).

Fontanon, Cl., G. Emptoz, J.-F. Archieri, Le fer dans l'enseignement du Conservatoire National des Arts et Métiers, in: F. Seitz (Ed.), Architecture et Métal en France. 19e–20e siècles, Paris 1994, p. 75–90.

Germann, G., Einführung in die Geschichte der Architekturtheorie, Darmstadt 1980.

Giedion, S., Building in France. Iron – Ferroconcrete, The Getty Center Publication Programs, Santa Monica, Ca., 1995 (first German edition: Leipzig/Berlin 1928).

'Le Globe. Recueil philosophique, politique et littéraire'; with respect to the founding of the École Centrale des Arts et Manufactures: Tome VI, N° 100 , Paris 8 October 1828 and N° 101, Paris 11 October 1828.

Grelon, A. and H. Stück (Ed.), Ingenieure in Frankreich, 1747–1990, Frankfurt a/M. and New York 1994; incl.: B. Kalaora and A. Savoye, Frédérique Le Play und die Le-Play-Ingenieure des 19. Jahrhunderts, p. 261–271.

Guillet, L., Cent ans de la vie de l'École Centrale des Arts et Manufactures 1829–1929, Paris 1929.
Hix, J., The Glass House, London 1974.
Hughes, Q., Liverpool, London 1969.
Jacomy, B., Die Société des Ingenieurs Civils auf der Suche nach einer Mission? In: A. Grelon, H. Stück (Ed.), Ingenieure in Frankreich, 1747–1990, Frankfurt a/M and New York 1994, p. 351–368.
Jenney, W.L.B. (with S. Loring), Principles and Practice of Architecture, Cleveland and Chicago 1869.
Jenney, W.L.B., Architecture. Lectures delivered at the University of Chicago, in: 'The Inland Architect and Builder', Vol. I, N° 2, March 1883 (Part I) up to Vol. III, April 1884 (Part X) [Archive and Library of the University of Illinois].
Jenney, W.L.B., A Few Practical Hints, Reprint and revised edition of Jenney's lecture for the Chicago Architectural Scetch Club of 28 January 1889, in: 'The Inland Architect and News Record', Vol. XIII, N° 9, February 1889, p. 7–9 [Archive and Library of the University of Illinois].
Jenney, W.L.B., An Age of Steel and Clay, Reprint and revised edition of Jenney's lecture for the Chicago Architectural Scetch Club of 6 October 1890, in: 'The Inland Architect and News Record', Vol. XVI, N° 7, December 1890, p. 75–77 [Archive and Library of the University of Illinois].
Jenney, W.L.B., Perspective View of Horticultural Building, World's Columbian Exposition, Chicago, in: 'The Inland Architect and News Record', Vol. XVII, N° 5, June 1891 [Archive and Library of the University of Illinois].
Jenney, W.L.B., Chicago Construction, or tall Buildings on a compressible Soil, in: 'The Engineering Record. Building Record & The Sanitary Engineer', Vol. 24, N° 24, New York, 14 November 1891, resp. London, 28 November 1891, p. 388–390.
Jenney, W.L.B., Autobiography of William Le Baron Jenney, Architect (written in 1906 for the Western Architect), in: 'The Western Architect', Vol. 10, N° 6, Minneapolis, Minn., June 1907, p. 59–66 [Archive and Library of the University of Illinois].
Lemoine, B., L'Architecture du fer. France: XIXe siècle, Seyssel 1986.
Mallet, M.A., Souvenirs et Mémoires. Autobiographie de J.-Daniel Colladon (Ed. Société des Ingénieurs Civils de France), Paris 1893.
Marrey, B., Le Fer à Paris. Architectures, Paris 1989.
Mary, Ch.-L., Cours d'Architecture, à l'École Centrale des Arts et Manufactures, Paris 1852–1853 (handwritten lecture manuscript and collection of plates; printed edition).
Le Moël, M. and R. Saint-Paul (Ed.), Le Conservatoire National des Arts et Métiers au coeur de Paris 1794–1994, Paris 1994.
Olmo, C., Il Lingotto 1915–1939. L'Architettura, L'Immagine, Il Lavoro, Torino 1994.
Perdonnet, A., Traité élémentaire des Chemins de fer, Tome I and II, Paris 1858, resp. 1860.
Peters. T.F., The Rise of the Skyscraper from the Ashes of Chicago, in: Invention & Technology, autumn 1987.
Peters, T.F., Building the Nineteenth Century, Cambridge, Mass., London 1996.
Picon, A. and M. Yvon, L'ingénieur artiste, Paris 1989.
Picon, A., L'invention de l'ingénieur moderne. L'École des Ponts et Chaussées 1747–1851, Paris 1992.
Pinet, G., Histoire de l'École Polytechnique, Paris 1887.
Poncetton, F., Eiffel. Le magicien du fer, Paris 1939.
Pothier, F., Histoire de l'École Centrale des Arts et Manufactures, Paris 1887.
Randall, F.A., History of the Development of Building Construction in Chicago, Urbana, Illinois, 1949.
'Rassegna', N° 28, Milano 1986: Perret: 25bis rue Franklin; incl.: P. Poitevin, La tecnica del progetto, p. 32ff.

Roisecco, G. (Ed.), L'architettura del ferro. La Francia (1715–1914), Roma 1973.
Salomon, G., Saint-Simon und der Sozialismus (Ed. P. Cassirer), Berlin 1919; with autobiography and a selection of essays.
Saint-Simon, Das industrielle System, 1821.
Saint-Simon, Der Katechismus der Industriellen, 1823/24.
Schild, E., Zwischen Glaspalast und Palais des Illusions. Form und Konstruktion im 19. Jahrhundert, Bauwelt Fundamente Vol. 20, Braunschweig/Wiesbaden 1983 (first edition 1967).
Seitz, F., L'Architecture du quotidien, in: M. Le Moël and R. Saint-Paul (Ed.), 1994, p. 182–184.
Seitz, F., L'École Spéciale d'Architecture, 1865–1930, Paris 1995.
Simmen, J. and U. Drepper, Der Fahrstuhl. Die Geschichte der vertikalen Eroberung, München 1984.
Singer, C. et al., A History of Technology; Vol. IV, The Industrial Revolution, 1750–1850, Oxford 1975 [4] (1958); Vol. V, The Late Nineteenth Century, 1850–1900, Oxford 1970 [4] (1958).
Steiner, F.H., French Iron Architecture, Ann Arbor, Mich., 1984 (first edition 1978).
Straub, H, Die Geschichte der Bauingenieurkunst. Ein Überblick von der Antike bis in die Neuzeit, Basel/Boston/Berlin 1992 (4th edition).
Sullivan, L., The Autobiography of an Idea, New York 1956 (first edition 1924).
Thomas, É., Histoire des Ateliers nationaux, Paris 1848.
Treue, W., K.-H. Manegold, Quellen zur Geschichte der industriellen Revolution, Göttingen, Frankfurt/M. and Zürich 1966.
Turak, Th., William Le Baron Jenney. A Pioneer of Modern Architecture, Ann Arbor, Mich., 1986.
Weiss, J.H., The Making of Technological Man, Cambridge, Mass., 1982.
Zola, É., Le Ventre de Paris, Paris 1964 (first edition 1873).

1
Lavoisier in his chemistry laboratory with his wife, Marie-Anne Lavoisier, who took part in his experiments, translated treatises and completed scientific drawings (painting by Jacques Louis David, 1788). (Nicole Dhombres, 1989, p.10)

2
The École de Mars on the Plaine des Sablons outside of Paris on the occasion of Robespierre's visit in July 1794. (Nicole Dhombres, 1989, p.88)

4
"Polytechniciens" as leaders of the July Revolution in 1830. (Callot, 1882, p.78)

3
Scholars and scientists with Napoleon under a tent during his expedition to Egypt. (Painting by Léon Cogniet, in: Napoleon 1802, p.16)

5
The glass pavilions in the Jardin des Plantes in Paris in the state it presents itself today: Replacements and reconstruction of the original greenhouses built by Rohault de Fleury in 1833. (photograph: the author, 1996)

6
Gare St. Lazare by Monet (1877); the tension rods of the "Polonceau Truss" were also apprehended by the artist. (Sagner-Düchting, 1994, p.94)

7
Lime Street Station built by Richard Turner, Liverpool 1849–1851 (the structure was later replaced and reinforced). (photograph: the author, 1989)

8
The office building "Oriel Chambers" on Water Street built by Peter Ellis, jr. in Liverpool (1864): facade facing the yard. (photograph: the author, 1989)

9
The Menier chocolate factory in Noisel-sur-Marne by Saulnier and Moisant (1871/72); the very first application of metal frame support structure with a curtain-wall facade attached to it. (Marrey, 1989, p.102)

10
Berlin Academy of Architecture by Schinkel (1832–1836). (photograph: W. Titzenthaler, around 1905, in Geist, 1993, fold-out in appendix)

11
Holabird & Roche (Gage Group, 30 and 24 South Michigan Ave., Chicago, 1898). (photograph: the author, 1991)

12
"The embodiment of Technology, in the form of a young woman" (relief of 1898 in the hall of the Conservatoire des Arts et Métiers, Paris). (Le Moël, 1994, p.139)

The Polytechnical Model in Austria

The establishment of the École Polytechnique (1794/95) and the École Centrale des Arts et Manufactures (1828/29) in Paris are exceptional pioneering accomplishments of school policy and educational science. Both institutions acted as driving forces for the rest of Europe – and also America – leading to the foundation of schools based on their example.

Prague and Vienna followed suit and set up polytechnical schools as first reactions to the new École Polytechnique in Paris. The "Bauakademie" founded by Frederick William III of Prussia in 1799 in Berlin adhered to a type of polytechnical program and subsequently formed one of the pillars of the later Technical College (Technische Hochshule). Additional polytechnical schools in Germany such as Karlsruhe were established after 1825.

Schools were also being founded at the same time in the United States of America, which first operated according to the polytechnical and then to the industrial model of education.

Italy and Great Britain followed suit in the 1860s and 1880s respectively. Italy's late entrance, in contrast to England, is to be seen within the context of its particular historical circumstances[1], as it did not form a political union until 1871 while Great Britain remained a leading industrial nation even without schools of engineering.[2] Furthermore, both countries' late development of scientifically based technical and industrial instruction for architects and engineers was no longer oriented on the original French models.

As in France, Austria also recognized as a consequence of the Enlightenment the importance of education as a motor for progress in technology, crafts and commercial industry. The promoters of scientific and technical education belonged to the commercial estates as well as to imperial commissions. Yet it was individual personages, such as Gerstner in Prague or Prechtl in Vienna, who stood at the front of this development and functioned as its standard-bearers.[3] Moreover, Austria had already developed a school tradition and a number of institutions upon which new

[1] "Legge Casati" of 1859; cf., also Guagnini, 1993, p.171–195, and Dadda, 1981.

[2] The importance of the "Great Exhibition"in Joseph Paxton's Crystal Palace of 1851 for the development of a systematic and industrially oriented education of architects and engineers in Great Britain will be treated further below (in a special chapter for this edition).

[3] Schoen, 1882, and Wickenden, 1929, p.46.

opposite page:
A comparison of the teaching programs of polytechnical schools of engineering in Vienna, Prague and Lemberg (around 1880). (Schoen, 1882, Appendix)

polytechnical schools were able to become established. Austria's orientation on French examples primarily occurred with respect to institutionalizing a systematic scientific and technical form of instruction on a higher level and, in the context of how the Enlightenment age understood it, on improving the quality of commerce and industry in order to place the nation's products at the service of its entire population.

The necessity of establishing a scientific and technological system of education in Austria from a chronicler's point of view. "Every one in the world regards himself as a technician, all kinds of people who possess everything but technical knowledge and who even less work in this particular profession take advantage of any opportunity to let their light shine, to advise on purely technical matters, to decide whether further technical advice is needed. (...) If the civil service is to work in the spirit of a government seeking the best as a blessing for the entire population, for the state and for the welfare of each individual, and if it is to provide these with the necessary protection all of the time and in every respect, swiftly and thoroughly, then this service must be organized appropriately and entrusted to the work and in the hands of a capable and for this profession specifically trained working force so that all accomplishments are completed according to legal regulations and with the participation of those versed in law.[4]

[4] Trans. from Schoen, 1882, p.6.

However, regional and national requirements stood in the way of a broader application of the French polytechnical model, which very quickly led to specialization in vocational education. However, certain disciplines, or subjects, continued to be taught throughout the curriculum that were not only geared towards specific professional branches. In this respect Austrian institutes of polytechnical education represent an *organizational transition* in the history of education; as a next step first Karlsruhe and then Zurich followed by installing comprehensive teaching disciplines or independent departments that in turn were adopted by other schools.[5]

[5] On the culture of education in the 18th century and on the forerunners of technical education in imperial Austria, cf. Schoen, 1882, p.7–9.

After the first polytechnical institute was founded in Prague (1806) similar schools followed in Graz (already in 1814), Vienna (1815), Cracow (1833/34), Brno (1843) and Lemberg (1844). Besides Prague and

B. Bauschule.

Lehrgegenstand und Uebung	Pflichtgegenstand der I. Staatsprüfung	Pflichtgegenstand der II. Staatsprüfung	Jahr	Wien	Winter-Semester	Sommer-Semester	Jahr	Prag (deutsch)	Winter-Semester	Sommer-Semester	Jahr	Prag (čechisch)	Winter-Semester	Sommer-Semester	Jahr	Lemberg	Winter-Semester	Sommer-Semester
				Wöchentliche Stundenzahl der Vorlesungen u. Übungen im														
Mathematik I. Curs	•		I.		7½	7½	I.		6	6	I.		7	7	I.		6	6
Correpetition über Mathematik									3	3			1	1	I.		3	3
Darstellende Geometrie	•		I.		4	4	I. II.	I. Curs II. Curs	3 2	3 2	I.		5	5	I.		6	6
Darstellende Geometrie, Constructives Zeichnen	•		I.		10	10	I. II.	zum I. Curs zum II. Curs	8 2	8 2	I.		10	10	I.		12	12
Propädeutik der Baukunst			I.		—	3	—								II.	Lehre der architekt. Formen	2	2
Architectonisches Zeichnen I. Th.	•		I.		6	6	II.		6	6	I.		2	4	II.		4	4
Architectonisches Zeichnen II. Th.			II.		10	10					II.		4	8	—			
Freihandzeichnen I. Curs	•		I.	Figurenzeichnen I. Curs	4	4	I.		12	12 (Hörer in Gruppen)	I.		8	8	I.		6	6
Physik	•		II.	allgem. u. techn.	5 —	5 —	I. II.	allgemeine technische prakt. Uebungen	5 2 2	5 2 2	I. — —	allgem. u. techn. Repetitorium	7 2	7 2	I. II.	allgem. u. techn. I. allgem. u. techn. II.	5 3	5 3
Elemente der reinen Mechanik			I.		4	4												
Mechanik	•		II.	techn. I. Theil	4	—	I.	I. Curs II. Curs	3 3	3 —	II. II.	Statik u. Dynamik Lehre üb. Elasticität und Festigkeit	5	5	II. II.	Uebungen	6 1	4 1
Mineralogie									—	—	I.	mineral. Practicum	6 1	— —	— —			
Praktische Geometrie (Niedere Geodäsie)	•			I. Theil	4½	4	II.		5	5	II.		6	6	II.	Geodäsie I. Curs	4	4
Praktische Geometrie, Situationszeichnen					3½	4	II.	Plan- u. Terrainz.	6	6	—		6	6	—		4	4
Praktische Geometrie Uebungen (durchschnittlich)					4	4		Feldübungen	4	4	—						2	2
Freihandzeichnen II. Curs			II.		4	4	II.		12	12 (Hörer in Gruppen)	II.		8	8	II.		6	6
Stereometrie									—	—	II.	Anfertig. v. Modellen	2 3	1 3	— —			
Hochbau		••	III.	Baumaterialien u. Bauconstructionen	6	6	III. IV.	I. Curs II. Curs	5 2	5 2	III. IV.	I. Curs II. Curs	6 3	6 3	III. IV.	I. Curs II. Curs	6 3	6 3
Hochbau, Constructionsübungen		••	III.		11	14	III.	zum I. Curs zum II. Curs	13 4	13 4	— —	zum I. Curs zum II. Curs	10 8	10 8	III. IV.	zum I. Curs zum II. Curs	12 18	12 18
Architekturgeschichte		••	III. IV.	des Alterthums des Mittelalters u. der Neuzeit	2 2	2 2	IV.		2 —	2 —	IV. V.	I. Curs II. Curs	1 1	1 1	IV. —		2	—
Baukunst		••	III. IV.	antike des Mittelalters und der Renaissance	3 3	3 3	IV. V.	Architektur I. C. Architektur II. C.	11 21	12 21	IV. IV.	Architektur II. Curs Architektur II. Curs	12 24	12 24	— —			
Architektonische Zeichnungs- und Compositionsübungen			III. IV. V.	I. Theil II. Theil III. Theil	8 12 9	8 12 9							—	—	V.		24	24
Ornamenten-Zeichnen		••	III. IV.	I. Curs II. Curs	6 6	6 6	III.		5	5	III. IV.		6	6	III.	I. Curs	6	8
Modelliren		••	III. IV.	I. Curs II. Curs	4 4	4 4	III.		6	6	IV.		6	6	—	II. Curs	6	8
Geologie		••	V.	I. Theil Gesteinslehre Uebungen	2 2	— —	III.	I. Curs	2	—	III.	I. und II. Curs Uebungen	— —	6 1	III.	Petrographie Uebungen	2 1	— —
Allgemeine Maschinenkunde		••	IV.		3	3	III.		5	—	III.		3	3	IV.		3	3
Allgemeine Maschinenkunde, Skizziren		••					III.	Repetitorium	4 1	4 1	— —		2	2	—			
Baumechanik		••	V.		—	5	III.		3 —	3 —	III. IV.	I. Curs (graph. Statik) Constructionen II. Curs Repetitorium	2 2 3 1	2 2 3 1	III. — — —	I. Curs Uebungen	4 4	— —
Mechanische Technologie											III.		4	4	IV.	empfohlen I. Curs	3	3
Encyklopädie der Chemie		••	IV.	anorgan. u. organ.	2	2	V.	anorgan. u. organ.	3	3	III.	organ. u. anorgan.	3	3	III.		3	3
Malerische Perspective			IV.		2	—	—											
Malerische Perspective, Übungen			IV.		—	4	—											
Encyklopädie des Strassen- und Wasserbaues		••	V.		3	—	IV.	Encyklopädie der Ingenieurbaukunde	— 2	— 2	IV. —	Encyklopädie des Ingenieurbauwesens	3	3	V. —	Encyklopädie der Ingen.-Wissenschaft	3	3
Encyklopädie des Brücken- und Eisenbahnbaues			V.		—	4	—	Skizzen	4	4	—							
Utilitäts-Baukunde			V.		3	3	—											
Utilitäts-Baukunde, Architektonische Compositions-Uebungen dazu			V.		6	6	—											
Eisenbahnhochbau			V.		3	—	—						—	—	V.	(empfohlen)		2
Eisenbahnhochbau, Compositions-Uebungen dazu			V.		4	4	—											4
Bau- u. Eisenbahngesetzkunde		••	V.		2		V.	Baugesetzkunde	2		V.	Baugesetze	2	—	IV.	Baugesetzkunde	– –	2
Buchhaltung									—	– –	V.		—	3	V.	(empfohlen)	2	2

Vienna only Lemberg and Cracow (until 1875) taught architecture or building design and construction. In the provinces teaching programs were more focused on regional demands.[6]

[6] Schoen, 1882, p.21–28.

The Polytechnical Institute in Prague

The beginnings of engineering education in Bohemia go back to the early 18th century. The first regular school of engineering was opened in 1718 and entailed Christian Joseph Willenberg's[7] lectures and practical exercises, a small collection of models used for demonstration purposes and a library. The actual teaching program was first instituted by Willenberg's successor Johann Ferdinand Schor after 1725. Besides offering basic subjects primary importance was also given to "civil architecture", bridge building and military architecture, drawing and practical exercises, excursions to construction sites, etc.. Among other things, Schor also offered the services of field and forest surveying and correction to the political estates.[8]

[7] Landscape engineer and founder of this school.

[8] Jelinek, 1856, p.3–19.

After Schor's death in 1767 Franz Leonhard Herget, Schor's student and assistant, succeeded him. Herget was able to substantially expand the teaching program and in his "Announcement" of 1769 he planned to create an early form of polytechnical instruction based on a broad spectrum of basic subjects such as arithmetic, geometry, trigonometry and others (to be accompanied by exercises), as well as a number of engineering sciences such as mechanics, statics, hydrostatics, hydraulics and, as a specific discipline, civil and military architecture, for which he also offered courses in drawing and design.[9] When the course of study on military architecture was taken over by the military academy and other military schools, the school's engineering science courses were assimilated into the department of philosophy of the University of Vienna after 1787. After Herget's death in 1800 the chair of engineering sciences had to be reorganized – representing the occasion for founding the polytechnical institute.[10]

[9] In the year 1770 165 students were enrolled in the school, in 1780 there were 213 students.

[10] Jelinek, 1856, p.19–26.

In the following years the idea became more clearly formulated to create an independent school of engineering to serve the commercial and industrial requirements of Bohemia and Austria-Hungary. Governmental infrastructure was provided around the turn of the century by district engineers who almost without exception were students of Herget. The plan to establish an industrial institution of higher learning in engineering was then taken up by Franz Joseph von Gerstner (1756–1832).

Gerstner studied at Prague University. At the age of 33 he was appointed by the univerisity as professor of mathematics. He was a pioneer of hydromechanics and as director of Bohemian waterworks (as of 1811) he planned and designed a canal link between the Moldova River and the Danube, which was rejected; from 1824 to 1832 he built in its place the first railroad on the continent between Budweis and Linz, though still operating with horse power! He wrote many scientific and technical treatises.[11] Already in 1797 he formulated a curriculum to be used as an example and as a guideline for future educational institutions of higher scientific and technical learning in Austria, and as of 1798 he held lectures at the "k.k. Studienrevisions-Hofcommission" (royal and imperial Court Commission on Study Inspections) "which aimed at none other than the establishment of a splendid technical educational institution according to the example of the famous Paris school of polytechnics."[12]

Franz Joseph von Gerstner, founder and first director of the polytechnical institute in Prague. (Jelinek, 1856, title page)

[11] Ricken, 1994, p.219; Straub, 1992, p.248.

[12] Trans. from Jelinek, 1856, p.26.

Gerstner's plan not only sought to retain and expand the subjects established by Herget as an "elementary course of physical and mathematical sciences" at the University's department of philosophy, but to also locate the "higher course or polytechnical studies" within an independent institute, a polytechnical school. According to Gerstner this program would comprise the subjects of natural history and minerals, chemistry and metals, mathematics and its applications, drawing, architecture and architectural engineering, mechanics and mechanical engineering.

Instruction was divided into lectures and practical exercises. As in the École Polytechnique instruction was held in classes, needy students received stipends, professors had assistants and at the end of the year exams were held. According to Gerstner's ideas the faculty body was also to form a "corps scientifique", which would become publicly active by participating in scientific conferences and through publications. The school was to serve the state and society in multifarious ways:

as a basic educational institution for teachers to improve commerce and industry, to produce scientific reports, to recruit civil servants and officers, etc. Jelinek, chronicler of the Prague polytechnical institute, at various points emphasized the fact that the Paris École Polytechnique was considered exemplary by Gerstner as well as the Court Commission.[13]

[13] Jelinek, 1856, p.30f.

In the process of discussing Gerstner's plan the Court as well as the estates emphasized the future polytechnical school's task to especially encourage national textile, glass and iron industry by installing two initial basic subjects, chemistry and mechanics; soon after architecture and architectural engineering as a third field was called for.

The school's inauguration and its first instruction classes began on 10 November 1806 in the St. Wenzel building. The school's goal was to "raise the commerce of the Fatherland through scientific instruction". It was to educate government civil servants as well as "factory managers" and engineers of private industry. After a few modifications the curriculum was divided into four parts: elementary mathematics and practical geometry, mechanics and hydraulics, agricultural and hydraulic architectural engineering – where instruction in general architecture and drawing was to took place – and general and specialized technical chemistry.[14]

[14] Jelinek, 1856, p.31–37.

As of 1812, after enrollment in this school had become obligatory for civil servants of government architecture and after the master builder title became contingent upon receiving a polytechnical diploma, student numbers at the school increased. As of 1817 industrially oriented special fields of instruction, such as the science of ironworks, salt chemistry or glass production were offered that stood in direct correlation to national industry. After 1820 the fields of agricultural economy and chemical mineralogy were also included. In 1822/23 the institute raised the art of drawing as a subject encompassing all disciplines.[15]

[15] Jelinek, 1856, p.39–56.

In the year 1820, challenged by the newly founded polytechnical institute in Vienna in 1815, Gerstner planned to expand his institute by assimilating a two-year basic training in the form of a Realschule (secondary modern school) upon which a three-year course of specialized study could be built upon.

In 1832, due to his advanced age, Gerstner retired and died in the same year.[16] In 1833/34 the modern secondary school (Realschule) inaugurated its courses with 60 students as part of the polytechnical institute. In 1839 a reorganization extended instruction in architecture and architectural engineering into a two-year course in order to include more recent building techniques such as chain bridges and railway engineering. In 1843 steps to reform the system of departments were undertaken by organizing the various fields of instruction, meaning its teaching program, in greater conformity to the various disciplines and professional fields.[17]

[16] Jelinek, 1856, p.56–71.

[17] Jelinek, 1856, p.76–85.

The political unrest of 1848 also had an impact on the Prague polytechnical institute. Obligatory examinations were subsequently dropped in favor of "frequentation reports". Even regular preparation provided at the Realschule level as well as the orderly transition into specialized study became modified. According to Jelinek, thanks to the director and the teaching faculty's strong leadership as well as to the student body's self discipline the school was able to continue its scientific activity. Because leading branches of industry such as the railway industry and continuing schools of mining actually required graduating diplomas for enrollment, the students' freedom to pursue their own program of study became relativized and, besides receiving frequentation reports, were thus forced in the end to pass the traditional final exams.[18]

[18] Jelinek, 1856, p.91–100.

As of 1863 instruction was given not only in German but also in Czech; in 1864 parallel classes were installed by language; in 1868 this development finally culminated in the establishment of two different and independent polytechnical institutes. In 1879 all polytechnical institutes of Austria were renamed and subsumed under the title Royal and Imperial Technical Colleges ("K.K. Technische Hochschulen").[19]

[19] Schoen, 1882, p.15.

The Prague school of polytechnics is not only the oldest one of its kind in Austria, but in the year of its founding in 1806 it could already look back on an almost hundred-year tradition in the polytechnical and enlightened concept of interconnecting science, the technical arts and commercial-industrial application within a polytechnical concept, which was oriented on the philosophy of the Enlightenment. In addition, it was the first school of engineering world-wide to base itself on the "original" model of the École Polytechnique. Prague therefore not only has the oldest university in Central Europe (founded in 1348) but it is also *the oldest technical school of higher learning in Europe outside of Paris.*

The Polytechnical Institute in Vienna

Vienna's polytechnical institute also had its precursors in history and was established according to specific regional and local conditions. Gerstner's curricular model of 1797, according to the Vienna polytechnical school's chronicler Joseph Neuwirth, is also to be considered as the starting point for Austria's own particular polytechnical development, which began in 1806 in Prague and was followed by Vienna in 1815.[20] Nevertheless, the manner in which the institute was founded and set up in Vienna diverged from that of Prague. First, the Parisian École Polytechnique was not primarily considered as an example worthy of being adopted to Viennese conditions; second, in contrast to Prague, Vienna was accustomed to academic freedoms and was confronted by existing institutes that already covered certain professional fields, especially architecture.

[20] Neuwirth, 1915.

From the very beginning these two aspects therefore also determined the discussion on the character and range of the new polytechnical school. Vienna was not interested in introducing "school discipline" or in "forcing" students to attend certain courses. The school was to be open to all those "artistically eager", and each was to determine his own course of study. In addition, graduating diplomas were not to be introduced in order to merely provide a system of confirmation. Hence, these principles dominated the school until its reorganization in 1863 when "obligatory instruction" was introduced.

The Vienna institute's actual starting point was in 1805 when the emperor set up a fund and ordered a report to be compiled on the Prague polytechnical school. As a consequence, different points of view emerged within the Court Chamber (which represented the estates, i.e. commercial interests) and within the Court Commission on Studies (Studienhofkommission) established as the school's founding body, which led to a ten-year deadlock. In 1810 Johann Joseph Prechtl was appointed to revise the project according to the results of the debate.

Prechtl (1778–1854), as the son of a manager of iron works under royal Würzburg control he was interested in issues of technology. As a 27–year-old he wrote a treatise "Über die Physik des Feuers" and as a result won the gold medal awarded by the Haarlem Academy of Sciences. He studied law and philosophy in Würzburg and subsequently entered Count Johann Taaffe's household in Brno as a private teacher, to whom he dedicated a paper titled "On the Mistakes of Education" (1804). In 1809 he was commissioned by the Austrian government to establish a modern secondary and navigation academy in Trieste. One year later he was appointed as teacher of physics, chemistry and natural history at the secondary modern academy (Realakademie) St. Anna in Vienna. In this position the government designated him to organize the new polytechnical institute. Prechtl directed the school for 35 years, was a member of the Academy of Sciences in Vienna and of many other academies and published 33 scientific and technical treatises and 90 articles in the "Technische Enzyklopädie" (Technical Encyclopedia), which he edited together with Altmütter and Karmarsch as of 1830.[21]

Johann Joseph Prechtl, founder and director of the Viennese polytechnical institute. (Neuwirth, 1915, p.220)

[21] Neuwirth, 1915, p.220f.

Joseph Wallis in his expertise summarizes the Prechtl plan of 1810 as follows: "The polytechnical institute is to accomplish the transition from pure theory to practice and is to teach the application of theoretical principles to individual branches of activity so that the processes upon which they are based can be introduced into the working place."[22] According to this the school was to contribute to overcoming the empirical method and ignorance of the workshops in the sense of the Enlightenment by providing a scientific and technical education. According to Prechtl the motivating factor behind this lay in the effort to concentrate national resources that had been consumed by the napoleonic wars, which had brought a loss of territory with them. Prechtl's curriculum of 1810 was goal-oriented, national and unique, meaning "Viennese", encompassing academic freedom and a model of instruction that was to maintain specialized education as well as a systematic and practical education in aesthetics.

[22] Wallis, trans. from and cit. in Neuwirth, 1915, p.14.

Prechtl did not intend to copy either the Parisian or Prague model. The three initial specialized disciplines were divided into chemical-technical, mathematical-technical and empirical-technical sections. This method of instruction was to combine theory and practice, testing and experimentation and demonstration by means of relevant models. Aesthetic education included art history, history of the arts and commerce, geometrical and model drawing, pattern and architectural design, and the theory of forms. However, this endeavor caused friction between the institute and the Academy of Fine Arts (Akademie der Bildenden Künste).[23]

[23] Neuwirth, 1915, p.20–50.

In 1814 Prechtl was appointed director of the future institute. In 1815, in the last moment before the opening, Prechtl traveled to Paris upon the invitation of the French Cabinet to view its technical institutions of education and – supplied with an imperial subsidy – to acquire chemical and physical apparatus, machine and pattern designs, publications and products of industry.

On 6 November 1815 the polytechnical institute opened its doors and on the following day Prechtl held his first lecture, which was attended by 47 students.[24] In his opening speech Prechtl delineated the tasks of the polytechnical institute in comparison to the Akademie der Bildenden Künste: "Are, then, the necessary or indispensable arts, which nevertheless form the very basis for important factories, manufacturers and technical industries in our state, not worthy of being given the same amount of attention?"[25] Prechtl's emphasis on the institute's equal importance and yet different goals paved the way for the later introduction of an independent course on building design and construction. However, as yet the school was to be divided between a technical and a commercial department.

[24] Neuwirth, 1915, p.51–64.

[25] Prechtl, trans. from and cit. in Neuwirth, 1915, p.66f.

The institute's scientific and technical program also entailed a collection, a workshop for mechanics, yearly exhibits on "national factory products" and the publication of a specialized school journal. According to Prechtl the polytechnical institute was to be a technical educational establishment, a museum of technology and an academy of technical sciences simultaneously.

The "Viennese Model". The Viennese model's "pièce de résistance" lay in its proclaimed academic freedom or the free provision of government-supported learning: "it was at least to allow each individual the possibility of freely taking advantage of this institute of instruction according to his respective talents and inclinations and according to his individual career choice; its admirable reasoning that 'school discipline and order can indeed exist without school compulsion' deserves a place of honor among the educational principles exhibited by any government office of that time."[26] In Germany this principle did not become established until polytechnical institutes were transformed into technical colleges as principles of higher scientific and technological education.

[26] Trans. from Neuwirth, 1915, p.58.

In this Viennese school's statement of purpose *an equal status with universities* is formulated for the very first time in the history of technical education: "The polytechnical institute will be for the commercially industrious bourgeois estates with respect to the practical arts and to the technical civil services that which universities initially are for educating civil servants and for the sciences as such."[27] This definition, which did not become a matter of course until half a century after it was formulated in the year 1816, precedes, for instance, the term "Sorbonne industrielle" coined as an assessment of itself by the École Centrale des Arts et Manufactures, founded in 1829.

[27] Trans. from Neuwirth, 1915, p.73f.

The polytechnical institute in Vienna, built 1816. (Neuwirth, 1915, p.3.)

Reports were given for completed courses and were accepted as official in the public sector as "factory authorization" and as qualifications for specialized professions such as for agricultural and district engineers. Like other polytechnical schools these reports were key certificates in attaining important permanent job positions.[28]

[28] Neuwirth, 1915, p.82.

The Vienna polytechnical school's technical museum was established as a "conservatory of arts and commerce" according to the Parisian example. Also, professorial travels as further training and as a means to "raise the national-industrial status historically" and the school's polytechnical journal with a concomitant professorial obligation to produce textbooks were also installed and followed the French model, intending to qualify its polytechnical members as carriers of a "society encouraging national industry." (Gesellschaft zur Aufmunterung der Nationalindustrie).[29]

[29] Trans. from Neuwirth, 1915, p.82f.

The struggle to establish a course on architecture was first fought on the level of rendering techniques. In 1827, ten years after the school was founded, the school of manufacturing and engraving design was reluctantly transferred from the Akademie to the polytechnical institute. In 1842 it was moved back, at which point the polytechnical school established its own design school.

In 1839 Prechtl installed in the field of agricultural and waterworks building construction a two-year course of study for building science and architecture and delineated the course's aim to provide "knowledge of building materials, the properties of which the architect must base his structures and joining elements on". At the same time it was not a matter of "copying but as much as developing one's own design as far as possible" so that "in manifesting the theories in individual objects the students would be given the opportunity to ponder and think upon them."[30] We see here, similar to Durand and Mary, that a *transitional theory of teaching and methodology from an academic to an industrial model of teaching with an emphasis on practical and useful process of design* has been developed, as well as model of independent learning and working.

[30] Trans. from Neuwirth, 1915, p. 129.

Subsequent to the far-reaching reorganization of 1863 resulting in a differentiated system of specialized schools within the polytechnical institute, building design and construction was established for the first time as an independent and complete course of study in 1866. It was first limited to teaching the theory of building material and building structure as well as architecture, architectural engineering and "the science of functional architecture". In teaching building materials the components of building design and construction and materials such as timber, stone and iron were treated. Already one year later the teaching program was augmented by general building theory and railway building construction. The latter became an independent discipline as of 1870 also attended by structural and civil engineers; as of 1897 it was fused together with functional architecture.[31] In 1875 the theory of architectonic forms was added, which included architectonic drawing, painterly perspective and a colloquium on urban planning. As of 1885 this school also appropriated art history. In 1894 heating, ventilation and firing technology was introduced.[32]

[31] Neuwirth, 1915, p.474.

[32] Neuwirth, 1915, p.497–541.

The polytechnical institute's reorganization of 1863 resulted in a system of specialized schools organized according to disciplines and profession the like of which had been developed by Karlsruhe and comprehensively introduced by the Swiss Federal Polytechnikum in Zurich in 1855. At this time "obligatory studies" were established and the students' unlimited freedom to determine their own program of study was abandoned. As of 1869 entrance exams were introduced and titled graduating diplomas were awarded, post-doctoral studies were offered and a change of name was decided upon. The organizational statute of 20 February 1875 formalized the transition from a polytechnical institute to a "Technical College" (Technische Hochschule). Finally, in 1880 a meeting of German-speaking technical colleges in Berlin decided to *complete the last step in acquiring equal university status*, i.e. the ability to award doctoral titles in the technical sciences.[33]

[33] Neuwirth, 1915, p.259–314.

One of the most prominent graduates of the Technische Hochschule of Vienna was, besides Camillo Sitte, Siegfried Giedion, who received his diploma as mechanical engineer in 1913. At that time mechanical engineering was intricately connected to other departments. The permeation of architecture, architectural engineering and technology throughout the field of mechanical engineering provided the fertile soil for Giedion's broad interests in science, technology and culture and for his interdisciplinary consciousness. Following his technical education he studied art history in Zurich and under Heinrich Wölfflin in Munich, under whom he also received his doctoral degree in 1922. As of 1938 he was professor at Harvard in Cambridge (Mass.) and lectured at the ETH (Swiss Federal Institute of Technology) in Zurich between 1947 to 1958.

On the Establishment of a System of Technical Higher Education in Germany

Because of Germany's quickly expanding industrial development a whole number of modern engineering schools came into being: Berlin (the Bauakademie in 1799 and Gewerbeakademie in 1821), Karlsruhe (1825), Darmstadt (1826), Munich (1827), Dresden (1828), Stuttgart (1829), Hanover (1831) and others. However, the first institution to provide education for scientific and technical engineers in Germany was Karlsruhe.[34]

[34] Bertrand, 1914, p.25.

Germany's delayed industrial development in comparison to Great Britain and France was due to its late dismantling of customs barriers and attempts at unification on the part of German federations of former principalities. A further factor that played a role was the country's constitutions (Prussia being the last to establish a constitution). A decisive moment occurred when a agreement was signed on 1 January 1834 between Prussia, South and Central Germany to establish the German customs union, or Zollverein. A Prussian constitution followed. The German federation of principalities remained intact, however, under Prussia's leadership. Towards the middle of the century national unity was not yet in place. (In this process the events of Paris in 1830 and 1848 influenced liberal-minded spirits in Germany.)

In France industrial development was from the very beginning closely connected to the establishment of a higher system of education in engineering. Although from early on in France commerce and industry were being promoted by the institutionalization of scientific and technical schools as well as the other way around, this did not occur in Germany until much later.[35] Furthermore, the educational task faced in Germany was a twofold one. First the population's level of education was to be raised in general and in particular also that of the common worker in order to stir and promote the understanding of the correlation between science, technology and industry on the one hand and social progress and the raising of living standards on the other.

[35] Lundgreen, 1973, p.139; cf. also id., 1975.

Moreover, a body of technicians was to be educated capable of mastering the high qualitative and quantitative, scientific and technical demands of Germany's surging industrial development. These tasks not only required that free entrepreneurship be promoted but that government also take the initiative: "Industrial entrepreneurship in favor of governmental interests therefore was predicated upon allowing businesses to take a greater part in government than heretofore. Thus, industrial development and constitutional progress were closely related to each other."[36] Shared government and business activity similar to that of French and Austrian societies and institutions is evident in, for instance, the Prussian "Association for the Promotion of Commercial Industry" (Verein zur Beförderung des Gewerbefleißes") founded in 1822, or also in museums, specialized journals, explorations and educational travels as well as in the organization of national and international exhibitions, and finally in efforts to make higher scientific and technical education available.

[36] Klemm, 1954, p.273.

The "Bauakademie" in Berlin

The establishment of the Berlin Academy of Architecture (Bauakademie) in the year 1799 was preceded by diverse developments. In this respect Frederick II founded the Freiberg School of Mining (Bergakademie) in 1765, offering systematic instruction as of 1770, and a school based on Enlightenment principles at his court in 1776, the "École de génie et d'architecture" based on the French model, which was, however, soon closed again. At the same time efforts were made to expand institutions of education such as the "Akademie der Bildenden Künste". Yet the civil service's growing demand for well-educated engineers could not be met by these institutes. As a reaction to this lack the royal Prussian Oberbaudepartementestablished the new Bauakademie.[37]

[37] Schnabel, 1925, p.19–21. See also: Schwarz, 1999.

The Bauakademie – A Kind of Polytechnical School. "The master builder, aware of the needs of architecture, hoped in vain for an institute that would transport in its instruction all branches of the art of building in their proper inter-relationship, and where theory and praxis go hand in hand in educating the prospective master builder. Especially important was this desire manifest in the intentions of the Prussian state, where in comparison so very much is built at royal expense and such great sums are spent on buildings of every kind."[38]

[38] Eytelwein, trans. from and cit. in: Schwarz, 1979, p.58.

From the very beginning the new Bauakademie was affiliated with the older Akademie der Bildenden Künste and was intermittently managed by the director of the art school or by the Oberbaudepartement. David Gilly taught the fields of bridge and hydraulic engineering, of melioration and port construction; his son Friedrich Gilly, who visited Paris in the years 1797 and 1798, took over the drawing course and Johann Albrecht Eytelwein, graduate of the Berlin Artillerieschule and first director of the Bauakademie, taught mechanical engineering.[39] At this time Friedrich Schinkel was studying at this school (1798–1800) and was especially influenced by Friedrich Gilly.[40]

[39] Print of the teaching program of 1 October 1799 in: Schwarz, 1979, p.76–77.

[40] Von Wolzogen, 1981, p.9.

Until the year 1824 the connection between the two schools improved in so far as handicraft apprentices and journeymen were also educated at the art academy and allowed to visit lectures at the building academy if they did well; and shared instruction was given in drawing and art history. The collaboration between the two academies ultimately found expression in Rudolf Schadow's directorship of both institutes simultaneously between 1816 and 1824. Under Christian Peter Beuth[41] as director the schools were separated as of 1824 and their courses of studies, even within the Bauakademie, became more differentiated. While the Bauakademie was to teach the technical aspects of architecture and building technology, the Akademie der Bildenden Künste was to take over "architectural theory". As a consequence of Schinkel's considerable influence at the Bauakademie this differentiation remained illusory.[42] Schinkel designed and erected the new Bauakademie building along the Spree canal himself (1832–1836; torn down 1961).[43] [color plate 10]

[41] Trade and commerce director of the Prussian ministry of finance.

[42] Schnabel, 1925, p.21f; Ricken, 1994, p.171ff.

[43] Geist, 1993.

The Berlin Gewerbeakademie

The development of technology and industry, of commerce and manufacturing forced the establishment of a new school type. In 1821 the Gewerbeakademie (Academy of Commerce) was founded by the State Council Kunth.[44] As of 1831 this school was considerably expanded by a technical dimension under Christian Peter Wilhelm Beuth. Public school education was a prerequisite for entering this school. Practical training was provided in one of the school's ancillary apprenticeship

[44] Reformer of trade and commerce as well as educator of the Humboldt brothers.

workshops. The school's aim was to educate skilled workers in theory and in science for commerce and industry as well as teachers for provincial trade schools, who in turn would educate those active in commerce. Beuth pursued the educational goal of creating skills and abilities that were to serve in "refining commerce through art". The teaching program ultimately maintained that all students were to attend the same basic scientific courses and that "they should attend all classes, achieve the same level in all subjects, and dedicate themselves to a career in chemical industry, in building technologies or in mechanical engineering."[45] This curriculum therefore exhibited rudimentary polytechnical principles and essentially left the student's later choice of career open.

[45] Trans. from and cit. in Lundgreen, 1994, p.175.

The union of both institutes (Bauakademie and Gewerbeakademie) within a "General School of Building" (Allgemeine Bauschule) under Beuth's directorship was unfortunately not to become permanent; Beuth's resignation in 1845 was followed by the schools' separation. A unified higher school of technical learning therefore was not yet able to become established in Berlin, and the long wished-for engineer educated in science and technology was to fall by the wayside.[46]

[46] Schnabel, 1925, p.23.

Plans for a Polytechnical School in Berlin

Parallel to the developments mentioned above, between 1820 and 1850 plans to establish a higher institution of education in mathematics and the natural sciences were also being discussed in Prussia's ministry of culture. Frederick William III's government heeded the critique provided by the visiting French inspector of higher learning institutions, Victor Cousin, and also obtained opinions on this issue from national scholars such as the mathematician C.G.J. Jacobi, the naturalist and explorer Alexander von Humboldt, and others, as efforts were being made among scientists, in engineering as well as in circles of trade, commerce, industry and in individual government ministries to rectify the lack of an École Polytechnique in Berlin.[47]

[47] Manegold, 1966, p.182f.

C.G.J. Jacobi "On the Paris Polytechnical School" (1835). In a lecture Jacobi elucidated the history of the Paris school, presented the formative individuals involved in developing its educational model and its organizational structure, explained the institution's intended relationship to continuing schools of specialization and closed his lecture by praising the Paris polytechnical school in the highest terms without ignoring to signalize the importance of such a school for Prussia: "The light that this polytechnical school radiated upon all of France through its teachers and through the achievements of its students has upheld it throughout all the storms of time because national honor participated in sustaining it. Soon mention of it on the platform of legislative power or in public documents was not made without calling it the first school in the world, the envy of Europe, an institute without comparison and without example."[48]

[48] Jacobi held his lecture in a public meeting of the physical-economic society of Königsberg on 22 May, 1835; trans. from Jacobi, 1891, p.367.

The Prussian ministry of culture's efforts to establish a polytechnical school of higher learning according to the French model were carried by the realization that future scientists, engineers and technical cadres – particularly also with respect to a strong Prussian government apparatus – must be educated and trained by an excellent school. An "amalgamation of natural sciences and technology with respect to commercial and productive industry" would be achieved as a result.[49] The intimate and inseparable connection between industrial progress and a higher system of education and training was therefore recognized. However, all attempts to found an independent polytechnical school in Berlin failed; in 1850 the project was finally abandoned.

[49] Manegold, 1966, p.184.

The Königliche Technische Hochschule in Berlin

In the 1870s Prussian government ministries and circles of building design and construction (architects and architectural engineers) again took up plans to unite the Bauakademie and the Gewerbeakademie. For this two different concepts – entailing a polytechnical school or a system of departments as exemplified by the polytechnical school in Zurich – were open to discussion. The decision to found a Royal Technical College (Königliche Technische Hochschule) in Berlin was made in 1876. While by 1875 the Bauakademie had completely separated the architecture department from engineering (road and hydraulic engineering) and instituted an additional specialized course for "private master builders"[50], in its plans for the technical college the government initially refused to separate architectural from engineering education.

[50] Ricken, 1994, p.172.

Nevertheless, when the Technische Hochschule opened its doors on 1 April 1879 it maintained five independent departments: architecture, civil engineering, mechanical engineering (including shipbuilding), chemistry and metallurgy, and the general sciences.[51]

[51] Schwarz, 1979, p.141–168.

Education and the History of Technology

Like the French schools a correlation exists between Berlin's Technical University and the history of German technological development. An example of this is the school's electrical engineering department: its creation was motivated by pioneers such as Werner von Siemens, Emil Rathenau and Adolf Slaby and as a result of its exceptional modern teaching and laboratory activity it played a major role in Berlin in establishing the city's electricity works, for instance. Siemens was educated in science and in technology at the Prussian United School of Artillery and Engineering ("Vereinigte Artillerie- und Ingenieurschule") and already in 1847 he established a telegraph company, which in 1887 became the world-wide company AEG.[52]

[52] Other examples are Hermann Rietschel, pioneer of heating and ventilation systems and first professor for this department as of 1885 at the Berlin Technical University, Franz Reuleaux, Alois Riedler and Georg Schlesinger, pioneers of the theory and construction of more complex, faster and larger machines, and Obergethmann and Nordmann, inventors of steam locomotives; cf. Schwarz, 1979, Part 6, p.277ff.

The list of Berlin's professors and also pioneers of the history of technology and architecture continues into the 20th century with, e.g. Konrad Zuse, inventor of the first programmed computer; Franz Dischinger, graduate of the Technical College (Technische Hochschule) in Karlsruhe, architectural engineer and pioneer of shell concrete construction and prestressed structures; and the teachers of architecture Hans Poelzig (as of 1923; studied at the Technische Universität in Berlin), Heinrich Tessenow (as of 1926; graduate of the Technische Hochschule of Munich) and Bruno Taut (as of 1930; studied under Theodor Fischer at the Technische Hochschule in Stuttgart).[53] One of the Bauakademie's most prominent graduates is Augustus Roebling (1806–1869). He emigrated to America in 1831, established a wire cable factory in 1840, designed many bridges including the suspension railway bridge at Niagara Falls in 1855, which spanned 250 meters, and the Brooklyn Bridge over New York's East River in 1869, which his son Washington Augustus completed.[54]

[53] Schwarz, 1979, p.277–462.

[54] His father died during the bridge's construction; Washington Augustus Roebling studied at the Rensselaer Polytechnic Institute; cf. Ricken, 1994, p.228f.

Joh. Augustus Roebling. (Steinman, 1945, p.132)

Finally, attention must be brought to the Technical University's annexation of the Royal Gardening Educational Institute (Königliche Gärtner-Lehranstalt), in

New York's Brooklyn Bridge by Roebling; original drawing. (Steinman, 1945, p.261)

1823, which was situated in Schöneberg and Potsdam and initiated by Peter Joseph Lenné, who directed it until 1865. From the very beginning Lenné included in his teaching program landscape, city and garden planning.[55] In 1811 Linné attended Durand's courses at the Paris École Polytechnique.

[55] Schwarz, 1979, p.117ff.

The Polytechnische Schule in Karlsruhe

Even though Berlin was finally able to create a new and unified technical college of polytechnical quality in 1879 by fusing the old Bauakademie with the more recent Gewerbeakademie, it was Karlsruhe that was to first "complete this development by becoming the first technical college on German soil."[56]

[56] Schnabel, 1925, p.23.

Forerunner Schools. The polytechnical college of the Fredericiana University in Karlsruhe can look back on a long tradition of scientific and technological education going as far back as 1779 when Professor Sander held his first seminar on "technology or the most noble crafts and arts" (in the winter semester) and on "economic natural theory" (in the summer semester) at the modern secondary school (Realschule) ancillary to the grammar school (Gymnasium) in Karlsruhe. This same educational establishment brought forth the "Physikalische Kabinett", which was the forerunner to the later institute of higher education in physics. The above mentioned efforts were a product of the Margrave Karl Frederick's enlightened spirit and of an era which was conducive to the development of "realistic" education.[57]

[57] Schnabel, 1925, p.23f.

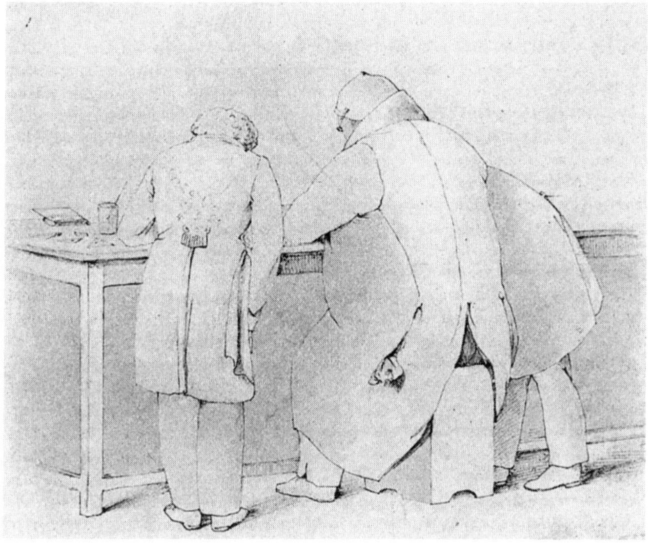

Weinbrenner and his students August and Paul Arnold. (Valdenaire, 1919, p.290)

A second institutional predecessor is to be seen in the "Architektonische Zeichenschule" founded in 1768 by the building director of Karlsruhe and builder of the its castle, von Kesslau. Instruction took place in the evenings, its students were craftsmen, masons, stone-cutters, carpenters, and metalworkers. The teaching program included projection drawing and copying, construction, cost calculation and the study of materials. However, "educating for industry" as a motto was not a welcome one among the masters here and as a consequence not many apprentices attended the classes. Not until the school was reorganized in 1796 by Weinbrenner, meanwhile under the directorship of Ch.F. Fahsolt, was this school able to be placed upon a new foundation in 1808 and increase its scope with 45 attending students. The school then became divided into departments.

Friedrich Weinbrenner's Bauschule. Weinbrenner (1766–1826) attended Fahsolt's drawing school and studied at the architecture academies in Vienna and Dresden (1790/91). As of 1797 he held the office as building inspector in Karlsruhe and as senior building director he exerted considerable influence, particularly after 1809, on Karlsruhe's architectonic and urban planning profile. Besides the older drawing school, or school for craftsmen, Weinbrenner established a "Bauschule" for architects which quickly became famous and was attended by many students outside of Baden; some of these students were, for instance, Joseph Berckmüller, Friedrich Dyckerhoff, Friedrich Eisenlohr, Friedrich Theodor Fischer, Hans Voss and many more. Weinbrenner's Bauschule subsequently formed the later polytechnical school's direct precursor, and in it "in fact the first section of the Technische Hochschule was already perfectly established."[58]

[58] Trans. from Schnabel, 1925, p.26.

These early schools – in their method of instruction rather than in their preferred and practiced style of building – already exhibited certain polytechnical principles such as instruction based on scientific knowledge, a combination of theory and practice, work in projects and designing in the form of exercises and summarizations compiled in student work groups. Furthermore, subjects were taught in a complete and standardized manner and with regard to practical application; hence, a systematic program and curriculum already existed. Moreover, elder students were also employed as tutors in exercise classes.[59]

[59] Valdenaire, 1919, p.291.

Reference to praxis not only occurred in practical exercises, in drawing and modeling classes but also in the courses of study in the sciences related to architecture: "A true architect must be able to use his head and hands equally well." On the other hand, the "school for the theoretical and practical master builder" must use exemplary works of architecture as material for illustration and comparison and must aid the student in applying his theoretical knowledge in practice, "possibly under the observation of a practical artist skilled in the subject."[60] Schirmer mentions many student works that show Weinbrenner's predilection towards assigning practical exercise work, especially in the area of building construction.[61]

[60] Trans. from Weinbrenner, preparatory notes on his architectonic textbook (1810), cit. in Schirmer, 1977, p.131.

[61] Schirmer, 1977, p.132; here predominance was given to timber construction and carpentry work, which Weinbrenner learned in his father's business.

A further aspect that anticipated polytechnical teaching was the primacy of the concept of function as the basis for architecture and its style that Weinbrenner emphasized in his textbook "Architektonisches Lehrbuch"[62] and that was related to Durand's theoretical approach. As Valdenaire, Weinbrenner's biographer writes, Weinbrenner's instruction was particularly characterized by his strong teaching personality exuding goodwill, warmth and a compelling manner of teaching that re-

[62] Valdenaire, 1919, p.284.

vealed enthusiasm for beautiful things and for artistic and cultural achievements in literature, painting and music: "Consequently, his institute brought forth not just one-sided experts and over-ambitious achievers but individuals of exceptional education and serious artists, as such they formed the nucleus of a new generation and continued to work in their master's footsteps."[63] From Weinbrenner's atelier activity and from the habit of his students to concern themselves with his ideas by copying their master's projects a veritable "Weinbrenner school" came into being.[64]

Friedrich Weinbrenner, sketched by Karl Sandhaas 1822. (Valdenaire, 1919, p.281)

Johann Gottfried Tulla's School of Engineering. In 1807 another school of engineering also opened in Karlsruhe. It was founded by the engineer who directed the Upper Rhine modification, Tulla (1770–1828). Like Weinbrenner, Tulla was trained in drawing at Fahsolt's school, in addition he studied at the Bergakademie in Freiberg between 1792 and 1794, and in 1801 he was commissioned by his sovereign prince to undertake a study trip to Paris. In France he was to analyze government facilities of hydraulic, dike and sluice building and subsequently apply these to Baden. Tulla also attended the École Polytechnique where he met Monge and practiced Géométrie descriptive (descriptive geometry). After his return in 1803 the Elector of Baden appointed him as senior engineer of the Rhine modification and also as professor of mathematics and reorganizer of Heidelberg University. This situation allowed him to combine theoretical teaching with practical work.

[63] Trans. from Valdenaire, 1919, p.293.

[64] Schirmer, 1977, p.133.

Johann Gottfried Tulla, painted around 1825. (Hotz, 1975, p.12)

Nevertheless, in a letter to the Elector of 15 July 1805 Tulla put forward the suggestion not to educate future engineers at the university but rather at a new institute in Karlsruhe. His considerations were based on the École Polytechnique model. Of Tulla's own invention was the establishment of "summer practical work" for students, who were employed at different sites each year "so that they gradually become acquainted with the various types of work and also acquire their necessary knowledge of the country."[65] This model of learning also went substantially beyond the "vacation work" (Travaux de vacances) introduced by the Parisian École Centrale founded in 1829.[66]

Tulla's plan was realized two years later in 1805 in the form of the first engineering school in Baden. In ap-

[65] Tulla, trans. from and cit. in: Schnabel, 1925, p.28.

[66] This early form of practical internship in construction and at sites was naturally focused on Tulla's principle building tasks.

pointing the mathematician and Tulla-student J.J. Ladomus to the school faculty not only was a general interest in pure mathematics to be encouraged among students but also their ability to "think for themselves" and their "inventive faculties"; this in turn reflects the polytechnical model upon which it is based: *With the help of teaching and learning methods combining theory and practice all the knowledge, skills and project experiences acquired were to be applied to every possible modern technical assignment.* As in the École Polytechnique the basic course in theory comprised three years of study, the students were between 13 and 16 years old and wore uniforms.[67]

[67] Schnabel, 1925, p.28 and 31.

On the history of establishing the Karlsruhe Polytechnikum. From 1808 efforts were made to unite Weinbrenner's Bauschule with Tulla's school of engineering, thus creating a polytechnical school. However, it was not until 1820 that the idea was realized to transform the modern secondary classes (Realklassen) affiliated with the Karlsruhe Lyceum (the former grammar school) into a higher polytechnical school as ancillary to the Physikalische Kabinett. In 1822 a ministerial decree ordered the immediate establishment of a polytechnical school, and thus the union of all polytechnical efforts. In this respect the foundation of the unified Polytechnikum in Karlsruhe on 7 October 1825 was an act of government dictated by the Grand Duke Ludwig.[68]

[68] Schnabel, 1925, p. 33.

The Polytechnical School's teaching program included according to grand-ducal decree not only mathematics and the natural sciences but also "their immediate, specialized application in civil activities of daily life (...)."[69] The courses were spread over three years: the first class was preparatory with a modern secondary (Realschule) character, the second treated applied mathematics and the third was titled "class of trade and commerce" (Handels- und Gewerbeklasse). After three years of basic education the students attended either the school of engineering, which was still directed by Tulla and placed under the department of engineering, or Weinbrenner's Bauschule where architecture was taught.[70]

[69] Grand-ducal Decree of 7 Oct. 1825, in: Bad. Staats- u. Regierungsblatt of 17 Oct. 1825, no. 23, trans. from and cit. in: Schnabel, 1925, p.33.

[70] Schnabel, 1925, p.33f.; Weinbrenner only taught at this polytechnical school for one year until his death in 1826.

In contrast to the situation in Berlin with its two separate schools the new Karlsruhe Polytechnikum offered not only basic education but also specialized training in imitation of the École Polytechnique (including its spectrum of subsequent schools of continuing education). Moreover, as in Paris mathematics played a key role in the school's methodology. One difference is that Karlsruhe took control of general preparatory education in the form of its ancillary Realschule, whereas Paris depended on Collèges and other similar schools. Another difference is to be seen in the standardized education for engineers not only engaged in civil service but also in private industry, which was to have a decisive influence on Karlsruhe's further development.

Nebenius' reorganization of 1832. After just a few years, already in 1832 the Polytechnikum underwent a fundamental reorganization similar to that of the Parisian École Centrale des Arts et Manufactures. This was undertaken by Karl Friedrich Nebenius.

Nebenius was known as the creator of Baden's early constitution of 1818. As a champion of dismantling internal tariffs (under his direction this had already been done within Baden in 1812), he represented the duchy in the Central German governments' negotiations for a German customs union. In 1825 he changed over from the ministry of finance to the ministry of the interior responsible for the new Polytechnical School. He then sought to reinforce the school's industrial orientation. In a book he defined the tasks faced by technical educational institutes as to contribute to raising production and to "prepare and facilitate a skilled management of production supplies" so that the expense of production could be decreased in relationship to the product results.[71] In this respect it was consistent for Nebenius to orient himself according to the example set by the new industrial, scientific and technical school of higher learning, the École Centrale in Paris.[72]

However in contrast to Paris, where each need was met by the establishment of a separate school, in Karlsruhe the classical model of polytechnical instruction was replaced by an industrial one within the institution itself. In further contrast to Paris, according to Nebenius this novel institute of education for engineers and architects was not to be a private enterprise. He particularly wanted the kind of high quality education offered in Paris to also benefit civil servants as well, as "the education of the higher productive classes is just as important to government as the competence of its civil servants."[73]

[71] Nebenius, Über technische Lehranstalten in ihrem Zusammenhang mit dem gesamten Unterrichtswesen und mit besonderer Riücksicht auf die Polytechnische Schule in Karlsruhe, (On technical institutes of teaching in their relationship to the entire system of instruction and with particular regard to the Polytechnical School in Karlsruhe), Karlsruhe 1833; cf. Schnabel, 1925, p.36–41.

[72] Schnabel, 1925, p.37.

[73] Nebenius, trans. from and cit. in: Schnabel, 1925, p.37.

The Polytechnical School's reorganization entailed separating the preparatory school (i.e. the first year course) and establishing it as a higher Realschule with the independent title of "Preparatory School" (Vorschule); it was still to be supervised by the Polytechnical School. As a second measure two mathematics classes were instituted to form the school's organizational core and as a third a broadened course of study was offered on the "practical level", i.e. continuing specialized schools were increased to five in which the traditional school of engineering and the Bauschule were further augmented by the school of forestry, trade and higher school of commerce (Höhere Gewerbeschule). The Bauschule also branched out into works management and architecture; the Höhere Gewerbeschule educated businessmen and managers and also followed the Tulla tradition.[74]

[74] Schnabel, 1925, p.41f.

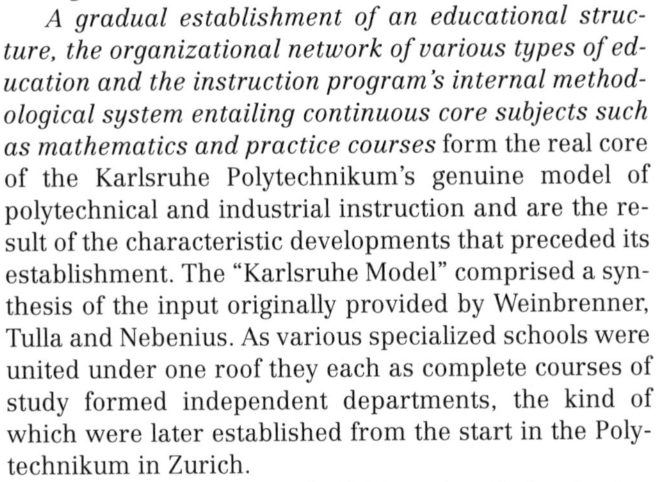

A gradual establishment of an educational structure, the organizational network of various types of education and the instruction program's internal methodological system entailing continuous core subjects such as mathematics and practice courses form the real core of the Karlsruhe Polytechnikum's genuine model of polytechnical and industrial instruction and are the result of the characteristic developments that preceded its establishment. The "Karlsruhe Model" comprised a synthesis of the input originally provided by Weinbrenner, Tulla and Nebenius. As various specialized schools were united under one roof they each as complete courses of study formed independent departments, the kind of which were later established from the start in the Polytechnikum in Zurich.

Reorganization as of 1841 under Redtenbacher. When Ferdinand Redtenbacher, who graduated from the Vienna Polytechnical School and subsequently taught at the Zurich Industrieschule ("School of Industry"), was appointed professor of mechanical engineering to Karlsruhe in 1841 a number of far-reaching reforms began.

Ferdinand Redtenbacher (1809–1863). (Hotz, 1975, p.32)

The Höhere Gewerbeschule in which he was to work was the least developed. Consequently, it was divided into a department of mechanical technology and of chemical technology. As of 1860 the former was designated by Redtenbacher as school of mechanical engineering and with him it became a well-known institute and "an independent domain for spreading the science of mechanical technology that he founded."[75] His methodic principle was again comparable to that of the École Centrale in that he combined theoretical instruction and the impartation of basic scientific knowledge with instruction in practical mechanical technology provided in workshops. Redtenbacher was responsible for establishing the Karlsruhe Polytechnikum as one of the leading schools of the world toward the middle of the century.[76]

[75] Trans. from Schnabel, 1925, p.47.

[76] More than 350 students were enrolled in the school, so that in 1859 a new building had to be erected.

In addition, Redtenbacher was a champion of "education for industry" and published many treatises, such as in 1840 on "Die polytechnische Schule" and its tasks. His conviction that technicians and engineers were not only to be educated in science and technology but that they also be provided with a humanistic education as means of integrating them into civil life and as factors in society's spiritual and intellectual development were soon realized by the creation of a chair for history at the Polytechnikum in Karlsruhe – a forerunner of the "General Studies Department" (Allgemeine Abteilung) at the Zurich Polytechnical School.[77]

[77] Schnabel, 1925, p.48f.

As of 1825 the Karlsruhe Polytechnikum provided the example for many others of its kind in Germany, throughout Europe and even America because it could already draw upon a long tradition and thus upon a greater amount of experience in organization and methodology; furthermore, the school was created from a synthesis of various educational approaches and expressed specific regional characteristics, thus acting not as an example to be exactly copied but as an inspirational model for integrating different characteristics and needs. Finally, the "Karlsruhe Model's" advantage lay in the fact that *it combined the different types exhibited by both predominant French models, thus creating a new, more mature, organizational and methodological model.*

The school constitution of 1865. Redtenbacher's successor was Franz Grashof who taught at the institute of commerce in Berlin before being employed by the school in Karlsruhe in 1864; he also founded the "Verein Deutscher Ingenieure VDI(German Association of Civil Engineers)" and edited its journal. In a crucial lecture given in Heidelberg in 1864, "On the principles of polytechnical schools that are to underlie their organization", in which he anticipated the development of polytechnical institutes into technical colleges, Redtenbacher stated the future task of polytechnical schools to unite the highest quality of education in science with professional technological training and emphasized the fundamental importance of mathematics, the natural sciences and the arts of rendering. These schools would prepare for work in civil service, private practice as well as in teaching.[78]

[78] Schnabel, 1925, p.53.

Again Karlsruhe set the pace and passed a new law of organization in 1865 that changed the polytechnical school into a technical college. On 20 January 1865 it received a university constitution that included appeal proceedings and self-administration and established its equal status to universities. It did not receive the title of "Technische Hochschule" until the year 1885.

An "Educational Motor" of the History of Science and Technology

Karlsruhe also brought forth a long and impressive list of scientists, engineers and architects who left their mark on the history of technology and building. One of these is the chemist Carl Engler (diploma in 1864), who was later appointed teacher to his former school, who founded a institute of technical inspection and experimentation and cooperated on creating BASF; Friedrich Hertz also taught in Karslruhe (1885–1889). Reinhard Baumeister studied in Hanover and Karlsruhe and instructed civil engineering at the Polytechnikum in Karlsruhe from 1862–1917, he founded the Verband Deutscher Architekten- und Bauingenieurvereine, which later became the Gesellschaft für Bauwesen ; Friedrich Engesser, founder of experimental statics, held the professorship for building construction, railway engineering and reinforced concrete building as of 1885.[79] (The Karlsruhe graduate and architectural engineer Maurice Koechlin has already been mentioned in the section on Gustave Eiffel; and Karl Culmann will be treated in the following chapter.) Some of the most renowned teachers of architecture after Weinbrenner were his students Heinrich Hübsch (professor between 1832–1854; as of 1842 Karlsruhe building director) and Friedrich Eisenlohr (teaching commission from 1832–1839, subsequently professor for structural theory until 1854); followed by Heinrich Lang (1852–1893),and Josef Durm (1868–1919)for architecture, and others.[80]

[79] Ricken, 1994, p.210–234; Hotz, 1975, p.47–52.

[80] In the 20th century the "Karlsruhe School" is represented for example by Otto Ernst Schweizer (studied at the Stuttgart Technical College and under Th. Fischer of the Munich Technical College, teacher for architecture and urban planning 1930–1961) and Egon Eiermann (studied under H. Poelzig at Berlin Technische Hochschule; professor of architecture at the Karlsruhe Technische Hochschule from 1947–1970); cf. Schirmer, 1975, p.15–128.

They all exhibited not only intense building but also committed teaching activity, yet they no longer followed a unified theory of teaching in architectural education so characteristic of Weinbrenner; their teaching also no longer represented a uniform approach to art as evident in Karlsruhe's building practice during Weinbrenner's time, in Schinkel's Berlin or in Klenze's Munich.[81]

[81] Schirmer, 1975, p.7f.

The System of Technical Higher Education and the History of Technology in Germany

Other polytechnical schools followed Karlsruhe's example and became technical colleges: Munich (1868), Dresden (1871), Brunswick (1877) and others.[82] As a result, not only were first steps taken in attaining an equal status with universities, leading to the acquisition of the privilege to bestow doctoral degrees as of 1899, but professional specialization also became introduced and the foundations laid for comprehensive research activity.[83]

[82] Schnabel, 1925, p. 53.

[83] Düwell, 1981, p.112.

Examples of early pioneers of industrial building history who had a polytechnical background include, for instance, Heinrich Gerber (1832–1912), who studied at the Munich Polytechnical School between 1849–1856 and submitted the first Bavarian patent for the "Gerber girder" in 1866; James Hobrecht (1825–1902), graduate of the Berlin Bauakademie (1847–1849) and from 1869 to 1892 chief engineer of the Berlin canalization and water drainage works; Johann Andreas Schubert (1809–1870), who after studying at the Bauschule of the Akademie der Bildenden Künste in Dresden was formatively involved as of 1828 in establishing the Königliche Technische Lehranstalt there (polytechnical school, later technical college). Many engineers did outstanding work in railway engineering and bridge building, or later in iron-reinforced construction and in theoretical treatises, textbooks and even building projects such as Georg Christoph Mertens (1843–1917) and Christian Otto Mohr (1835–1917), graduates of the Polytechnical School in Hanover, the latter appointed to the Polytechnical School in Stuttgart in 1867 and in 1873 to Dresden, August Föppl (1854–1924), student of Mohr at the Stuttgart Technische Hochschule, and many more.[84] Other prominent graduates of Stuttgart include Charles Louis Strobel[85] and the architect Jakob Friedrich Wanner, builder of Zurich's main railroad station (1867).

[84] Ricken, 1994, p.210.234.

[85] Strobel came from Cincinnati in the United States, graduated in 1873 in Stuttgart; in the 1880s he became a specialist in the method of steel-frame construction and was employed as such for the Carnegie Group in Chicago; cf. Peters, 1996, p.266.

In conclusion, the Karlsruhe school's influence on Switzerland is to be mentioned. Already before the Polytechnikum was founded in Karslruhe, Konrad Escher von der Linth called in Tulla to work on the major Linth modification between the Zürichsee and Walensee (1807–1816), who had been a pioneer in directing the Rhine modification.[86] The manifold organizational and biographical cross-references that run between the polytechnical schools of Karlsruhe and Zurich will be treated in more detail further on.

[86] Straub, 1992, p.247f.

Extended wing of the Karlsruhe Polytechnikum's principle building around 1880. (Hotz, 1975, p.36)

The Swiss Federal Institute of Technology's principle building in Zurich. (Niggli, 1930, annex, illustrations)

The Beginning of Education in Architecture and Engineering in Switzerland

The leading institute of higher learning for engineers and architects towards the middle of the century and directly after the French educational institutes was the "Eidgenössische Polytechnikum" in Zurich, founded in 1855 at the time the Parisian École Centrale was brought under government control. As a result of the particular conditions existing in Switzerland with regard to the political union of diverse cultural environments and in consideration of the trends exhibited in Germany and France, the Polytechnikum developed a unique program and model of instruction that not only referred to the École Centrale des Arts et Manufactures (however not to the École Polytechnique) but – in its initial program – also to Karlsruhe.

Establishing a higher institute of technical learning in Switzerland proceeded along various stages in which also international experience could be drawn upon and adapted according to national conditions. The first Swiss institutions to be established as a part of this development were the Military School in Thun and the polytechnical department of the University of Lausanne. Like everywhere else, preparatory instruction for continuing technical education was carried out on the intermediate level of education and out of the internal necessity to boost industrial development, raise it to a scientific niveau and to stir interest in the professional fields of architecture, engineering and technology. In this respect is particularly the "Industrieschule" in functioned acted as of 1855 as a preparatory school to the Swiss Federal Polytechnikum.[87]

[87] Bandle, 1994, p.9–19.

Forerunners of the "Polytechnikum"

As already mentioned, Dufour, a former student of the École Polytechnique (diploma 1807) and graduate of the École du Génie (diploma 1810), founded in 1819 the Military School in Thun based on the French example. A further "Polytechnicien", Pierre-Joseph Marguet, and two other "Centraux", Jules Marguet and Louis Rivier, were formatively involved in the inception of the "École polytechnique de l'Université de Lausanne", which opened in 1853 under the name "École spéciale de Lausanne", was renamed in 1869/70 into "Faculté technique de l'Académie de Lausanne" and again in 1890 into "École d'ingénieurs de l'Université de Lausanne"; it became independent of the university and is today known as the École Polytechnique Fédérale de Lausanne (Swiss Federal Insitute of Technology, Lausanne), an affiliate of the ETH in Zurich.[88] Both Jules Marguet and Louis Rivier received their diplomas from the École Centrale, the former as "Constructeur" in 1840 and the latter in chemistry in 1843.[89] The course of study first comprised two, then three years of education. The teaching program was almost a verbatim version of that of the École Centrale's and particularly encompassed the applied and industrial sciences, including the planning and design of iron bridges and railway facilities, also of public building of representation, etc.; this makes evident that the education of architects had already begun there. The novel course on structure was taught by Pierre-Joseph Marguet.[90] The polytechnical school in Lausanne was the first Swiss institute of technology to offer complete courses of study, thus allowing young Swiss students to be educated in their own country.

[88] Paschoud, 1953, p.15–25.

[89] Of the five who founded the school were also Jean Gay, rector and professor of the Académie de Lausanne and Henri Bischoff, likewise professor of the academy.

[90] Pierre-Joseph Marguet, after attending the École Polytechnique, went on to the École des Ponts et Chaussées and held the office as senior engineer of the French Corps des Ponts et Chaussées; in this function he was responsible for the Departement Pas-de-Calais, were he supervised the construction of the Boulogne-sur-Mer port facilities and the light towers of that region.

On the Events Leading up to the "Eidgenössische Polytechnikum" in Zurich

Since 1798 when Napoleon invaded Switzerland, the country had been confronted with a concept of engineer and officer education as established by the French model. The idea was taken up in the same year by Philipp Albert Stapfer from Brugg, Helvetian minister of education, by submitting a proposal to create a Swiss university or central academy which would offer the advantages of a German university (scientific niveau, academic freedoms to determine one's own teaching methods and freedom of the student to determine his own program of study) and combine it with those of the Parisian École Polytechnique; with this aim in mind he wrote a Helvetian Board "Direktorium" message dated 18 November 1798.[91] The formation of a polytechnical university, or as Stapfer called it, a "policy school" was put off over a long period of time due to subsequent military and political events, at least while the Cantons' re-

[91] Oechsli, 1905, p.3.

lationship to central government remained unclear. In 1803 the Cantons once again were granted the power to control education and higher systems of learning.

Until the middle of the century focus was maintained on projects relating to universities or university-like institutions, which nevertheless were coupled with proposals to also establish chairs for technology, agriculture and forestry or architecture, road, bridge and canal construction or military affairs; however, none of these proposals were realized. Even a "polytechnical department" as part of the University of Bern did not go beyond the planning stage.[92]

[92] Oechsli, 1905, p.18f.

On the other hand, a system of technical education did establish itself "from below" similar to developments in Germany, this time via art and "Realschulen" and via schools of commerce and industry. Here the initiative was taken by committed individuals of science and education, technology, industry and politics. In this way the first industrial schools were founded in Zurich, Winterthur, Bern and Lucerne, and from a historical perspective they are to be considered as preliminary stages leading to the establishment of the "Eidgenössische Polytechnikum" in Zurich.[93]

[93] Oechsli, 1905, p.19–33.

The articles of education in Switzerland's federal constitution. General Dufour's successful campaign in the Sonderbund War of 1847 not only heralded national unity and the Federal Constitution of 1848 but also reawakened the idea of a Swiss school of higher learning as a symbol of that unity. In the context of the "Kommission für Revision des Bundesvertrags" (Commission to Revise the Federal Contract) summoned on 17 February 1848, a project proposal was submitted and discussed whether a polytechnical school should also be established parallel to a national university. After the final debate that took place between the 14 and 24 June 1848 the Cantons finally agreed upon an optional provision to be included in the "Tagsatzung" (Federal Assembly): "The Federation is authorized to establish a university and a polytechnical school."[94] This version was included as Article 22 in the Federal Constitution of 12 September 1848.

[94] Federal Constitution of 12 September 1848, Art. 22, trans. from and cit. in: Oechsli, 1905, p.56.

The commission on higher education of 1851. The choice of location for the new Swiss school of higher learning was connected to the issue surrounding the site of the new federal capital: since Bern was to become the capital of Switzerland, Zurich was to take over the school of higher learning. Stefano Franscini from the Canton of Ticino, who had already been an early proponent of a central institute for higher learning, was the first to hold the office of interior minister. On 30 April 1851 he presented to the Federal Council a comprehensive statistical study on the state of higher education[95], which proved a great need for and lack of places of higher learning in Switzerland.[96]

Already in 1849 a commission on higher learning was established in order to implement Article 22 of the constitution. Its members included, among others, Guillaume-Henry Dufour (Geneva), Alfred Escher (administrative president, Zurich), Alexander Schweizer (rector of the University of Zurich), Professor Peter Merian (Basle); Federal Council Franscini presided. The discussions proved Escher to be the actual "spiritus rector" behind Swiss plans to establish institutes of higher learning. On 29 May 1849 the commission voted with a majority to erect a university as well as a polytechnical school; however, the university was given a priority.

While Dufour worked out a syllabus in a subcommittee based on the example provided by France's two major schools, the École Polytechnique and the École Centrale, Escher took a different direction: the Karlsruhe model. The chronicler Oechsli says the following about Dufour's plan: "Yet it was less the militarily organized 'École Polytechnique', predominantly charged with preparing students for military and civil service with its boarding school and subsequent continuing schools, than the 'École Centrale des arts et manufactures' founded in 1829 to fulfill the needs of French big industry and until 1851 remaining a private, yet state-supported school that Switzerland deemed worthy of copying."[97]

The commission's deliberations resulted in substantial changes to the original polytechnical plan. Consequently, Escher was given the task to undertake the final editing of the comprehensive articled statute, and he in turn appointed a specialist, Joseph Wolfgang v. Deschwanden, "who was subsequently to become the polytechnical school's actual coordinator."[98]

[95] Between 1846 and 1848, 652 Swiss nationals studied at foreign schools of higher learning, of which 96 attended polytechnical schools; in the year 1849, 25 studied in Karlsruhe, 11 in Paris (École Centrale as well as the École Polytechnique), including 8 in Munich and Vienna each.

[96] Oechsli, 1905, p.58–64.

[97] Oechsli, 1905, p.70.

Joseph v. Deschwanden (1819–1866). (Oechsli, 1905, p.120)

[98] Deschwanden was a student of Redtenbacher (who was responsible for reorganizing the Karlsruhe Polytechnical School as of 1841) at the Zurich "Industrieschule", he completed practical work at the Escher, Wyss & Co. Machine Works, studied at the polytechnical schools of Karlsruhe and Stuttgart and in 1847 became rector of the "Industrieschule". During this time he undertook study travels to France and Great Britain and visited the "Great Exhibition" in London (1851); cf. Oechsli, 1905, p.71.

According to Escher's plan the Polytechnikum was to continue from schools of industry and commerce and include a department of civil engineering, mechanical engineering and chemistry each; the age of 18 was stipulated as the minimum age of entrance and German and French prescribed as the major languages. Escher's scheme also delineated a methodological instrumentarium comprising a system of repetiteurs, excursions, competitions and regular theoretical and practical examinations; here the promoters of polytechnical education were also able to draw upon the French experience. Professors were elected to life positions, paid and given pensions by the Federal Council.

Alfred Escher (1819–1882). (Guggenbühl, 1955, p.203)

Escher's plan did not prescribe a standardized course of study with certain specialization for all students similar to that of the École Centrale or as imagined by Dufour, but rather in imitation of the model provided by Karlsruhe it provided education in a specific professional field: "The direct influence of French examples clearly evident [in Dufour's proposal; the author] is not as pronounced in Escher's; instead, that of German institutes are detectable, especially that of Karlsruhe, which at that time was unique in its strict arrangement of specialized schools reminiscent of a university system of departments."[99]

[99] Oechsli, 1905, p.75.

Because v. Deschwanden, as advisor to Escher on this issue was well acquainted with Karlsruhe and stood in personal contact to Redtenbacher, the emphasis on that particular school form is to be attributed to him. In contrast to Karlsruhe, the "Zurich Model's" initial proposal offered only civil engineering, mechanical engineering and chemistry and neglected Karlsruhe's school of construction, forestry and trade; military education as demanded by the National Council was to remain in the responsibility of the school in Thun and not transferred to the polytechnical school. Escher's subsequent design ultimately accommodated 10 professorships, including one for architecture, and provided for national stipends.

Stefano Franscini (1796–1857): relief located in the Swiss Federal Institute of Technology's (ETH) principle building which was given to the school on the occasion of its 100th aniversary by the "Gesellschaft ehemaliger Polytechniker"(GEP/ Group Lugano) ("Schweizerische Bauzeitung", special issue 1955, p.607)

The "Polytechnical School Law" of 1854. Between 27 June and 1 July 1851 the commission on higher learning's final deliberations took place. Escher's report supported by v. Deschwanden's expertise was accepted without opposition. In the deliberations in both chambers of the Federal Assembly of 1854 the commission's proposal to locate the university and the polytechnical school in the German and French part of Switzerland respectively and the Federal Council's suggestions (to dispense with stipends, introduce lectures on war economy and national economy in all departments) were contested; and the "Polytechnical School Law" of 7 February 1854 was finally settled. On 22 April Zurich accepted the Federal Assembly's "College Bequest".[100]

[100] Oechsli, 1905, p.81–122.

Teaching Program and the Inauguration of the "Eidgenössische Polytechnikum"

Immediately after the law of 7 February 1854 the Federal Council commissioned an expert committee to work out requirements and "regulations for a Swiss polytechnical school" in which once again Escher and v. Deschwanden participated. In their concluding report submitted to the Federal Council they proposed to substantially increase the courses of study to be offered from three to five: schools architecture (Bauschule), civil engineering, mechanical technology, chemistry and forestry. A "school of higher education in exact, political and humanistic sciences" prescribed by the law was to encompass a sixth department as a "department of philosophy and national commerce and industry". Student attendance of the latter was to be voluntary and organized instruction chosen according to individual interests and inclination.

The "Zurich model" is therefor to be considered a further development of the example provided by Karlsruhe; it offered complete and independent courses of studies oriented towards specific fields of profession and enhanced the importance of general education for architects and engineers. The sixth department of general education was divided into three areas: the natural sciences, mathematical sciences and subjects pertaining to literature, national industry and commerce. The courses offered here were of a basic and comprehensive nature and included, for instance, theoretical and experimental chemistry, a subject the Parisian École Centrale termed as "Science industrielle"; mathematics also included astronomy, and the third group entailed among other things four languages, history, art history, free-hand drawing.[101]

[101] Oechsli, 1905, p.152.

Innovations instituted by Zurich particularly worthy of being mentioned are the *"prize competitions"* which served to "encourage a scientific environment among students" and were held in three of the six departments alternately entailing a first and second prize each; the names of those awarded prizes were to be published in the official federal journal. A novelty was also *the administration of exams at the end of each school year and diploma examinations.* The school of architecture bestowed "master-builder" diplomas. Meanwhile the interconnection of theoretical instruction with practical exercises, project work and laboratory exercises had become a established part of polytechnical and industrial education in architecture and engineering.

Carl Culmann (1821–1881). (Oechsli, 1905, p.176)

On 7 July 1854 the Federal Council decided in favor of the proposals submitted by the expert committee under Escher and v. Deschwanden and designated the inception of the Zurich Polytechnikum for October 1855.[102]

[102] On the opening celebrations and the opening speeches, cf. Wolf, 1880.

The School and Teaching Goals Existing During the "Poly's" Founding Era. "The Polytechnical School endows Swiss industry with leaders and officers enabling it to successfully compete in the fierce international market; it also provides government with technically educated administrative personnel so necessary to accomplishing the modern tasks facing a state; it produces the country's architects of monumental edifices, its engineers to drill mountain tunnels, dam rivers and build bridges, its foresters responsible for reforesting Switzerland's mountain slopes and for neutralizing the danger posed by mountain torrents, its teachers of agriculture, of intermediate schools and not least of the Polytechnical School itself; in short, there is possibly no area of life upon which it does not exert a positive influence."[103] How much of the original "Esprit polytechnicien" is expressed here in this chronicle of 1905, coupled with the spirit of a henceforth modern and industrial, nationally minded Switzerland also open to the world, sentiments that express all the hope and energy invested in the country's first federal institution of higher learning!

[103] Oechsli, 1905, trans. from and cit. in: Niggli, Zurich, 1930.

Albert de Mousson (1805–1890). (Oechsli, 1905, p.188)

Jakob Burckhardt (1818–1897). (Oechsli, 1905, p.208)

[104] De Mousson was teacher of mathematics at the Industrieschule in Zurich from 1833 until his appointment to the Polytechnical School and also worked as full professor at the University of Zurich.

[105] Oechsli, 1905, p.171–122; an overview of the professors in the first phase is also to be found in: Guggenbühl, 1955, p.70–72.

[106] On the cultural, intellectual and scientific background concerning professors appointed from Germany, France and Italy, cf.: "Schweizerische Hochschulzeitung", 1955, p.37–58.

[107] Guggenbühl, 1955, p.94.

The First Teaching Faculty

The Federal Council subsequently advertised all professorial and teaching positions in Swiss, German and French newspapers and with regard to the school's inauguration appointed a number of internationally respected professors to teach at the new "Zurich Polytechnikum". These including Gottfried Semper for the school of architecture, the railway engineer and graduate of the Karlsruhe Polytechnikum Karl Culmann for the department of civil engineering, Alexander Bolley for chemistry and Albert de Mousson, "Polytechnicien" (diploma 1827) for experimental physics.[104] The chair for mineralogy was to be held by Arnold Escher von der Linth, son of Hans Konrad Escher, coordinator of the Linth modification; the field of descriptive geometry ("Géométrie descriptive") was given to v. Deschwanden, and Franz Reuleaux, who studied in Karlsruhe and subsequently worked in Berlin and Bonn, was appointed by the school to the department of mechanical technology. Friedrich Theodor Vischer from Tübingen was designated to teach German literature, Francesco De Sanctis Italian literature, and for art history possibly the most famous specialist of art and antiquity was called to teach art history: Jakob Burckhardt from Basle.[105]

The first teaching faculty applied itself with great commitment and organized class instruction. Until World War I foreign professors predominated; in this context the school admitted many refugees from the German revolution of 1847/48, including Semper, Bolley and Vischer.[106] Around 1880 an almost complete turn-over of the entire faculty occured when Burckhardt soon followed a call to Basle, Semper to Vienna (1871) and even Vischer and Mousson resigned; Bolley, v. Deschwanden and Escher von der Linth had passed away in the meantime.[107]

Preparatory Course and 1855 Inception; The Beginning Phase

In regard to the regular course a preparatory course was installed according to the example provided by Karlsruhe after Nebenius' reform (1832) in which v. Deschwanden, de Mousson and others were involved. The first entrance exams were held on the 23rd and

24th of April 1855 in the University of Zurich's building. The preparatory course of instruction was given from the 1st of May to the 28th of September of that year; 50 successful candidates from all over Switzerland subsequently enrolled in the Polytechnikum.[108]

October 15, 1855 was designated the opening date of the Polytechnical School. The Federal Council appointed v. Deschwanden as the school's first rector. Semper, Ferdinand and Julius Stadler, Culmann and v. Deschwanden taught at the school of architecture. Semper also was the first person to hold the office of departmental head of that school.

The Students. The number of students enrolled in the school increased dramatically.[109] During the school year of 1858/59 109 regular students were registered at the school, five years later it had already increased to 560 and in the school year 1875/76 the number of students attending the school exceeded the one thousand mark with 725 "polytechnicians" and 289 non-official attendants. In this early period the Polytechnikum took over as the leading international institute and attracted many foreign students to Zurich, as remarked by Urner in the Festschrift on the school's 125th anniversary: "Ever since the Polytechnikum has presented itself in full splendor in the Semper Building it has been a major force of attraction to Southern and Northern Germans, to Austrians, to students from Hungary and a few other countries. Among matriculated students, those of Swiss nationality posed but a minority for a decade."[110]

Women in higher technical education. The first female student to attend the Zurich Polytechnikum was Nadina Smetzky from Moscow; she was accepted to the school of mechanical technology (department of mechanical engineering) in 1871/72.[111] Other female students attending other departments were either Russians or Germans. The first Swiss female polytechnical student was Maja Knecht from Zurich, who graduated in 1895 as a specialized teacher in natural sciences.[112] The first woman to receive a diploma from the Poly was a Russian named Marie Kowalik from Chernikov, namely in 1877 from the department of agriculture and forestry.[113] Nevertheless, the number of female students enrolled during that time remained relatively small.[114]

[108] The preparatory course's aim was similar to that of the "Cours préliminaires" offered by the École Polytechnique in 1794/95 and improved the accessibility to higher education; it continued to be offered until 1881.

[109] Bergier, 1980, p.657ff.

[110] Urner, 1980, p.29.

[111] School News from the 1871/72 school year in: Programm der eidg. Polytechnischen Schule für das Schuljahr 1872/73, Zurich 1872.

[112] Schweiz. Verband der Akademikerinnen (trans.: Union of Swiss Female Academics), 1928, p.54.

[113] School News from the 1876/77 school year, in: Programm der eidg. Polytechnischen Schule für das Schuljahr 1877/78, Zurich 1877.

[114] Guggenbühl, 1955, p.97f.

In contrast to this, women were not allowed to enroll in the École Centrale in Paris until the year 1917[115], and a 1970 law allowed women to finally register in the École Polytechnique.[116] Considering the active role women played in France in the context of philosophy and the Enlightenment it is astonishing how delayed this development was. In 1898 the Conservatoire des Arts et Métiers dedicated a symbolic relief to women as the embodiment of "La technique". [color plate 12]

Like in other polytechnical and industrial schools of higher learning, graduates of the Zurich Polytechnikum endeavored to establish an alumni association; in this respect a first meeting of the "Association of Former Students of the Swiss Federal Polytechnical School" was called in 1869 and established in 1878 (later called the Association of Former Polytechnicians, "Gesellschaft Ehemaliger Polytechniker", GEP). This association aimed at "creating and maintaining friendly relations between members of different graduating classes, promoting professional efforts and the technological sciences."[117] Like their "Polytechniciens" or "Centraux" kinsmen, Swiss "Polytechniker" also pursued a three-fold aim of amicable contacts, support and sociability, subject-specific scientific discourse and the promotion of science and technology – goals that are not confined within national borders. From this a worldly "Esprit polytechnicien" also developed in Zurich, which is still maintained and in effect to date. The association of former Swiss Polytechnicians also supported a publication: "Eisenbahn – Chemin de Fer: Bulletin Polytechnique – Zeitschrift für Bau- und Verkehrswesen", which was also an organ of the still existing Schweizerischer Ingenieur- und Architekten-Verein (Swiss Association of Engineers and Architects).[118]

The Building Issue. The greatest initial problem was posed by the question of where the institute was to be housed. The school appropriated rooms in the university as well as in the Industrieschule, which made its laboratory available to the institute. In addition, among other the former St. Augustine Monastery, the seminary administration in the Kirchgasse, and the Kappelerhof were appropriated for use. In 1857, in the context of planning the erection of a new building, its future was decided upon. The Canton subse-

[115] École Centrale, 1981, p.75.

[116] In 1972 seven women passed the entrance exams for the first time, one of which was Vietnamese. In addition, one of these first "Polytechniciennes", Anne Chopinet, easily took first place among all accepted candidates for that year. The number of women students subsequently attending the École Polytechnique remained small and wavered between 10 and 20 until 1980; the "Polytechniciennes" were just as obliged to complete military service as their male counterparts and were subject to just as much discipline. Already in 1922 students of the school disseminated the idea to accept women students and also proposed an appropriate uniform; cf. Callot, 1982, p.192f. As a point of interest, at the end of the 19th century French higher schools of chemistry, e.g. in Marseille, Rouen, Rennes and Strasbourg institutionalized mixed classes from the very beginning, and in 1925 a "École Polytechnique Féminine" was even founded in Paris; cf. Marry, 1994, p.239.

[117] Guggenbühl, 1955, p.98.

[118] The first shared issue came out in 1876.

The Polytechnical School's Central Building Hall by Semper; after an original drawing by G.Gull. (Niggli, 1930, appendix, illustrations)

quently launched an architectural competition, which closed in 1858 and attracted 19 designs from Switzerland, Germany and France; however, all designs were declared unsuitable (Semper was a member of the Jury). As a consequence the School Council appointed the task to work out the building program directly to Semper.[119] The new building was erected at its current location in the year 1864.[120] Additional building extensions were designed for the department of chemistry (1884–1886) and of physics (1887–1890) and were directed by the architecture professors Bluntschli and Lasius.

[119] Guggenbühl, 1955, p.87.

[120] Oechsli, 1905, p.225–258; on the building's development cf. especially Niggli, 1930, p.25–95, and the plans and photographs in the appendix.

School reorganizations and curricular revisions. After the first reform was instituted in 1866 the school of architecture was renamed into "College of Building Design and Construction", was furnished with assistants and the right to confer the title "qualified master builder or architect"; moreover, a new course of studies, a teacher training institute in mathematics and the natural sciences, was established as a 6th department, thereby transforming non-specific studies into a 7th department. In 1869 the forestry department was converted into a department of agriculture and forestry, in 1874 the Polytechnikum established a military school under its roof for the first time and in 1880 the "Swiss Federal Material Testing Institute" (Eigenössische Materialprüfungsanstalt, EMPA) was founded as an ancillary, annexed institute.[121] In the year 1881 the School Council passed new entrance qualifications which prescribed preparatory intermediate schools to institute six and a half years of basic education in order guarantee direct enrollment to their students without entrance exams.[122] In 1899 a revision of the school's regulations resulted in the establishment of an 8th department comprising an independent section on military sciences; a further revision of 1908/09 transformed obligatory programs of study into normal programs of study, establishing a certain amount of academic freedom in determining individual study programs, and in 1911 the Polytechnikum received its ultimate title of "Eidgenössische Technische Hochschule" (Swiss Federal Institute of Technology) and the privilege of conferring doctoral degrees.[123]

[121] Guggenbühl, 1955, p.90–93.

[122] Guggenbühl, 1955, p.103.

[123] Urner, 1980, p.34.

The "Poly" in the Industrial Era

The Zurich Polytechnikum was founded at a point in time that marked the transition from the "pioneering era" to the "second industrial era" (Gründerzeit).[124] The predominant branches of the economy were initially the most rich in tradition, such as the textile industry – particularly the cotton mills – that had been undergoing a process of industrialization since 1801, and the mechanical industry, which through the pioneering efforts of Hans Caspar Escher and Jakob Sulzer-Neuffert could claim equal international representation since the 1820s with respect to the production of

[124] Bergier, 1990, p.202–261.

tool and textile machinery, turbines, pumps, machines for processing wood and manufacturing paper, etc.. This was followed by the chemical industry in the area around Basle; yet although the seeds were laid in the 1860s, major enterprises did not develop that could match German competition until the 1880s (CIBA, Geigy, Sandoz).

The way towards industrialization was paved by the establishment of a federal constitutional government (1848), Swiss customs tariffs and the standardization of coinage (both in 1850), and was facilitated by the first international trade agreements, first with Great Britain (1855) and France (1865), then with

"The Aluminum Industry Joint-Stock Company Chippis and its research institute in Neuhausen at the Rhine Falls convey their best wishes to the Swiss Federal Institute of Technology on their hundred-year anniversary." Greetings advertised by the Aluminum Industry on the occasion of the Swiss Federal Institute of Technology's hundredth-year anniversary in 1955 published in the special issue of the "Schweizerische Hochschulzeitung" (p.149).

Austria-Hungary and Italy (1869) and finally with the German Zollverein (Customs Union; 1870).

Again, like everywhere else, the railway provided the impetus for industrial development and its internationalization. Although the legal foundations for a national railway network had already been laid in 1850 (Law of Expropriation) and had found exceptional support in the person of Alfred Escher, who was president of the National Council in 1849, Switzerland witnessed a "train delay".[125] Escher himself established the Swiss Credit Bank in 1855 and as an instrument of investment it made a advance towards the establishment of the "Northeast Railway". By establishing further companies a railway network was able to encompass all of the Swiss Midland around 1865.

[125] Bergier, 1990, p.307–323.

After the construction of the Brenner railway in Austria (1867) and the Mont-Cenis tunnel connection between France and Italy (1871), the Gotthard railway was the next major European project (1871–1880); it was placed under Escher's supervision. The project's building contractor, Louis Favre from Geneva, collaborated with different engineers, one of them being Colladon[126] who helped the former to substantially accelerate the drilling process by his invention of special water-cooled and pneumatically powered compressors as of 1878. As a next project the Simplon tunnel was inaugurated in 1906.

[126] Teacher at the École Centrale des Arts et Manufactures between 1829 and 1836.; see also Mathys, 1955.

It was within this economic and industrial context that the tasks faced by graduates of "Poly" in Zurich lay – moreover, the school became closely linked with industry, particularly from the very beginning.

Engineers working at the vanguard of technological innovation and industrial development were educated at the department of mechanical engineering[127] and civil engineers were responsible for accomplishing projects of infrastructure, such as bridge, road and railway construction. With Culmann – the first teacher at the department of civil engineering who was educated in Metz and Karlsruhe and who published seminal work on "Graphic Statics", laying the foundations for its use as an exact science and thus initiating a shift in paradigm in theory and practice – a veritable "*Culmann School*" was created that brought forth also prominent pioneers of e.g. bridge building such as

[127] On the "Poly's" interconnection with Swiss mechanical industry, cf. Oechsli, 1905, Vol.2, p.449ff.

Wilhelm Ritter, Maurice Koechlin, Robert Maillart and many more, and that is still effective today (Christian Menn and his students).[128]

[128] On Culmann, cf. Maurer, 1998.

Semper's School of Architecture
The "School of Architecture's" educational program for architects entailed basic courses in the first year, e.g. design and construction, structural and material theory, drawing and ornamentation, theory of style, descriptive geometry and projection, stone- and wood-cutting, mathematics and natural sciences, clay and plaster modeling, woodworking; the second year reinforced instruction in aesthetics with courses on civil and public building and the theory of building styles, architectural drawing, design projects in road, railway, bridge and waterworks construction. In addition it offered classes in the theory of mechanics and mechanical engineering, geology, jurisprudence, figure and landscape drawing and modeling. Finally, in the third year the more creative aspects of architectural activity were stressed: civil and monumental building, larger projects, painterly perspective and ornamentation design, archaeology, history of architecture, figure and landscape drawing and modeling.[129] The school of architecture aimed at educating master builders in technology and art for "civil and monumental architecture".

Gottfried Semper (1803–1879). (Oechsli, 1905, p.172)

[129] Oechsli, 1905, p.148f.

The "Semper School". Graduates of Semper's school of architecture were therefore, in accordance with the school program, predominantly involved in representational building. The aim was to erect administrative buildings for the new national state in municipalities, cities, Cantons and on the federal level, then postal buildings and processing halls for the railway and finally hotels, banks, schools, hospitals, etc. Within the context of the "Semper School", from which approximately 100 students graduated between 1855 and 1870[130], 35 were involved in public building and hotel construction, 17 in the railway business (railway stations, administration), 20 took on public offices (departments of planning and building construction on all levels, political offices), of which 2 were even Federal Councils[131]; 16 became teachers after studying, partly in industrial and technical schools, at uni-

[130] Only those are meant who attended all of Sempers's three-year course of study; the report follows Fröhlich, 1974, p.181–239.

[131] Joseph Anton Schobinger as of 1908 and Frédéric Louis Perrier, Schobinger's successor as of 1912.

versities and schools of higher learning, including Semper's own successor at the Polytechnikum, Alfred Friedrich Bluntschli (as of 1881). Some became painters, others became businessmen (all branches are represented).[132]

Semper's stay in London: the industrial context. Gottfried Semper (1803–1879) profited particularly from his exile in London – which was forced on him for his participation in the failed Dresden Revolt of 1849 – in that he on the one hand gained practical demonstrative knowledge of modern standardized iron and glass technology by designing four national pavilions in the 1851 "Crystal Palace" (Canada, Turkey, Sweden and Denmark) and by his close contact to Henry Cole and Joseph Paxton[133]; on the other hand, under Cole he assumed tasks of teaching and methodology by teaching metalworking (1852–1855) and other subjects at the "Department of Practical Arts". At this newly established institute in South Kensington, London, which was to push ahead a reform of "design schools" after 1851, subject-based classes were introduced for the very first time and Semper was the only teacher with experience available, gained from his time in Dresden.[134]

Based on his initial industrial observances and contacts in London, already in October 1851 Semper wrote a paper on "Science, Industry and Art" where he proposes an industrially oriented program of education for the commercial and industrial application of the arts as well as for the field of architecture and engineering. In his paper he bases his suggestions on the example provided by the Conservatoire des Arts et Métiers in Paris and even refers to the current lectures offered there for the winter semester of 1851/52. In addition, Semper proposes the introduction of workshops and the organization of class competitions as instruments of teaching and methodology. Collections of illustrative material was to be used in the form of goods produced from art, technology, industry and architecture.

Semper's theory was based on the assumption that teaching "comparative building theory" must unite at its core of methodology all other crafts and technologies, products and processes of ceramics, textile pro-

[132] What is also revealing are the cross-references in the educational background of Semper's students. In this respect as many students changed their place of study to Zurich (e.g. 6 from Karlsruhe, others from Stuttgart, Munich, Hanover, Vienna, etc.) as those transferring from Zurich to other continuing schools abroad (of which 13 went to Paris, mostly to the École des Beaux-Arts, 10 to Berlin, 8 to Stuttgart, and 6 to Vienna); Fröhlich, 1974, p.254f.

[133] Semper made further efforts to work under Paxton; Herrmann, 1978, p.32–71.

[134] Cf. Emmerson, 1973, p.166ff.; Giedion, 1948, p.348–360; Reising, 1976, p.51f.; Herrmann, 1978, p.71–93.

duction, woodworking as well as masonry and engineering as industrially applied arts: "the interplay of these four elements under the authority of architecture."[135] His experience as a teacher in London was followed by a treatise on a modern "Plan of instruction for the section on metal and furniture technology" in 1852/53 and also a "Report on the Department of architecture, metal and furniture technology and practical design". (1853/54).[136]

[135] Semper, 1852; cf. also Wingler, 1966, p.62–68; see also Lankheit, 1976, p.44f.

[136] Semper, 1853, and Semper, 1854; cf. also Wingler, 1966, p.83ff, and 87ff.

Semper's ideas on the organization of instruction, teaching and learning are not only formulated in these papers but the contents conveyed by his lectures are also carried by the Beaux-Arts tradition: "Those who study follow the master of the atelier in his practical work. In this way they come into some contact with praxis and are given the best opportunity to acquire practical sense and experience."[137] This "teaching relationship" is characterized by a dependence on the "Maître", who encourages learning through imitation and sets the standards. In lectures and exercises binding references are established to exemplary works from antiquity through history (in chronological order) from which a "theory of style" or theory of building types is created. Yet Semper is also aware of the advantages posed by a commercially or industrially oriented school: "It seemed necessary to me for all students to be united within one room in order to enjoy the advantages of a system of mutual encouragement and mutual instruction, the kind of which I have experienced to produce the most effective motivation in art classes."[138]

[137] Trans. from Semper, 1853, p.101; cf. also Wingler, 1966, p.83.

[138] Trans. from Semper, 1854, p.99; cf. also Wingler, 1966, p.90.

Teaching Activity in Zurich: Architecture and "Style". When Semper was appointed to the Zurich Polytechnikum in 1854/55 he could draw upon a great deal of experience in teaching, beginning in Dresden where he could closely combine his teaching activity with building practice (1834–1849), then in London, where he was exceedingly challenged by the industrial culture there. Besides his experiences in building practice in Great Britain, Semper came into increasing contact with industrial construction through his continuous contact with, among others, Hittorff and Gau in Paris, which he had maintained since his years of learning there (1826/27).

Nevertheless, the ideas he formulated in total point in two very different directions: his interest in industry on the one hand, which are manifested in the papers, theories and teaching of his London years, and on the other hand his building practice in stone and his classical architectural theory. In his teaching at the Polytechnikum the aim and function of a building as well as references to material in architectonic design were superimposed by issues of "style".[139] He taught his students of architecture the following: "True monumental substance is the rectangular parallelepiped building [Quaderbau]."[140] However, in his lectures in Zurich Semper used modern illustrative material and even presented examples from iron engineering as manuscripts and college notebooks prove.[141]

Iron in Architecture. It is astonishing that the "Semper School" emerged at the same time as the protagonists of modern architectural engineering were being educated at the École de Centrale in Paris: Eiffel (1855), Jenney (1856), Moisant (1859) and Contamin (1860). In the expert discourse of the time and on the important issue concerning the use of iron or steel and metal in architecture, in his writings Semper took on a cautious and skeptical position as well as one that exceptionally presages modern architectural theory between the two World Wars, whereas in his teachings, in semester projects and in his own building practice or in those of his students hardly any of this becomes evident.

Again and again Semper referred to iron as a building material, first in his "Iron Structures" (1849) and at last in the closing chapter of "Style" on "Metallotechnik". He was first able to accept iron for use in buildings "of definitive practical intention, for protective roofs that covered wide spans, especially for railway station halls", yet goes on to criticize the Bibliothèque Ste. de Geneviève for which "M. Labroust thought it right to erect an unfortunately visible iron roof structure and on top of that to cover it in dark green paint so that the library hall, which is also the reading room, fails to exude an atmosphere of comfortable isolation so necessary to serious study and which no one will hardly exit from with pleasurable feelings." Semper prefers using iron in lattice and

[139] On Semper's teaching activity in Zurich (instruction in design, instruments of teaching and methodology, the form and goals of instruction in design, study schedules, competition assignments, excursions organized by the school of architecture and student works), cf. in detail Fröhlich, 1974, p.76–102.

[140] Trans. from and cit. in Herrmann, 1981, p.68.

[141] In this context, for instance, he presented the Bibliothèque de Ste. Geneviève by Labrouste, as well as railway stations, markets and halls and exhibition buildings as comparisons between London and Paris; Herrmann, 1981, p.141ff.

mesh work and as rods, "yet not as girders for large masses, as building support, or as basic melody of the motif." He subsequently asks himself how metal is to become part of "the beautiful building craft" and then goes on to express the idea motivated from "calculation and experience" that hollow metal prisms and metal worked out as metal sheets be used as structural elements: "But as far as I know, no one as yet has brought forth this structural method architectonically. I am of the opinion that this must happen if art is to appropriate a share of iron." Here Semper points to the iron roof structures of his Museum project in Dresden designed as a system of metal sheets and draws the conclusion: "Metal (...) only in sheet form (under this term I understand every form which in proportion to its mean area has a large surface; I also include in this for example hollow cast columns) is capable of being used in the art of building."[142]

[142] Trans. from Semper, 1849, p.485–487; cf. also Wingler, 1966, p.22–24.

Here Semper's theoretical approach makes a great leap into the future, consider for instance the experiments undertaken in deforming, folding, arching and forming sheet metal begun by Jean Prouvé as of 1925 in building facades. Semper also opens up the problem of tectonics to the building material of iron, an issue he treated in detail in his opus Magnum – "Style".

Already in his "second major work" ("Style", Vol.1) he states the following on the importance of iron and metal: "Among the substances that man uses for his purposes, metal is the one which combines all of the above characteristics of raw materials; its form can be smelted and hardened, it is flexible, hard, capable of resisting wear and tear to a high degree, is very elastic and of considerable relative strength (...)", etc.. Because of iron's manifold possibilities of treatment and the diverse technical processes that can be applied to it, Semper considers it necessary to "dedicate a special section on Metallotechnik to it."[143]

[143] Trans. from Semper, 1860, p.12.

In this section, which forms the last section of his major work "Style" ("Eleventh Chapter. Metallotechnik in volume 2), it again seems necessary to Semper to "first consider the last mentioned process of expansion, straining and bending, etc." (p.479). In accord with his basic cultural anthropological approach, Semper begins with ornamentation, masks and pro-

Department store Jelmoli by Stadler und Usteri in the city of Zurich (1899). (Oechsli, 1905, vol. 2, p.416)

tective apparel, dividing the latter into systems of fabric, scales, rings and hollow sheets (p.485). In bringing together Semper's concepts of material and technological processing with his tectonic principle of "opposing forces of active and passive elements of a structure, of settings and surroundings, of ornamenting and ornamentation" (p.502) it becomes easily possible to draw a connection to the development of modern building technology of the 20th century, at least to structures with hollow molds and formed metal sheet, and even to the actual debate on tectonics.

However, Semper warns of the transference of this progressive thought to representative architecture – and here is where the root of his approach to teaching at the "Poly" in Zurich is to be seen: "The dangerous idea that from iron structures applied to monumental buildings a new style of building must emerge has already led astray a number of architects, who though talented, were alienated from high art." (p.550) At the close of the book Semper mentions useful examples of the application of cast-iron, though only as examples without being exemplary: "The exhibition building in Paris. The new one in London. Bibliothèque de Ste. Geneviève, Paris.

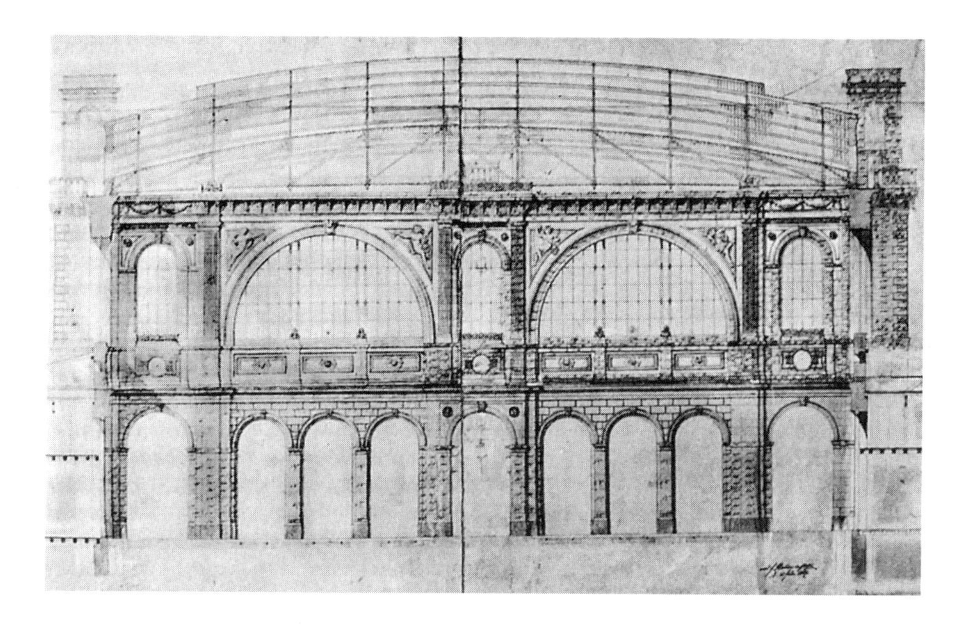

Cross-section of the station and reception hall of Zurich's Main Railway Station; Working plan by Jakob Friedrich Wanner (1867). (Stutz, 1983, p.173)

The roof structures of many railway stations" (p.588).[144] Metal, as prophetically recognized by Semper in its possibilities, is a raw material of tectonics and as such is subordinate to style in architecture.

Industrial Building Assignments in Switzerland of the 19th Century

The "primate of style" as exemplified by the "Semper School" as of 1855 is, however, also partly due to a lack of major industrial building assignments (market halls, department stores, galleries, exhibition pavilions, etc.). Semper and his graduates were concerned with representative buildings for federal, Canton and municipal institutions. Of the few modern buildings of steel and glass on still exists, the Jelmoli department store in Zurich city (architectural firm Stadler & Usteri, 1899).

The few major track-covering halls erected as railway stations for the Swiss railway companies, which were not built until they became state-owned in 1872, either directly involved only the activity of engineering offices or metal construction companies rather than architects, e.g. the engineer L. Bösch in Winterthur (1895), the firm Bell AG Kriens in Lucerne (1896), Buss AG in Basle (1907), the firm MAN for the Badischer Bahnhof in the German part of Basle (reception build-

[144] Trans. from Semper, 1863, p.478–589. On Semper's position on iron as a building material, cf. also Herrmann, 1981, p.61–68.

ing: Karl Moser; 1913), or Buss AG also for the railway hall in Lausanne (1916).[145] Only in one instance, in the construction of the Zurich train station, was the architect Jakob Friedrich Wanner to collaborate with the engineers Karl Culmann and Friedrich Seitz (assembled in 1867/68 by the Nuremberg firm Klett).[146]

In the early years of railway engineering the following buildings were constructed in Switzerland, among others: the entrance hall in Lucerne (1859) and the station halls of the Swiss Central Railway in Bern, Thun and Basle (1860) that were planned and designed by the railway architect Rudolf Ludwig Maring; the distances were still spanned by means of a combination of wood and iron structures. An example of the cooperation between architects and engineers is also Romanshorn (Johann Jakob Breitinger with August Beckh, 1855) and also St. Gallen (Felix Wilhelm Kubly with Friedrich Bitzer, graduate of the Stuttgart Polytechnikum, 1856).[147]

Besides station halls for railways major production and assembly halls of machine and metal industry and other factories posed a challenge to builders; here too particularly engineers and businesses led the way, not architects. Thus, Semper and his graduates were to concentrate on erecting representational buildings, for example in the important industrial city of Winterthur (Semper's council office building, town hall arcade by Joseph Bösch[148]).

The School of Architecture According to Semper

Semper remained in Zurich only until 1871 when he was called to Vienna. His successor was Friedrich Bluntschli, who worked there from 1881 to 1914. The school subsequently appointed Gustav Gull (1900–1929) and later introduced a new course on architecture based on the "architecture of modernity" taught by the group of teachers Karl Moser (1915–1928), Otto Rudolf Salvisberg (1929–1940) and Hans Hofman (1941–1957) as well as William Dunkel (1929–1959).[149]

Nevertheless, Semper's principles of teaching and methodology established in the architecture course was to remain intact for the next one hundred years – and a "gate house" still comprised the introductory subject in first-semester architecture until 1956![150]

[145] Stutz, 1983, p. 180–253.

[146] In the planning phase, Wanner, graduate of the Stuttgart Polytechnikum (studied under Christian Friedrich Leins), undertook a few trips to foreign countries included Paris more than once, also in the year 1863, and visited the Gare du Nord by Hittorff and Couche still under construction.; he returned to Paris once again in 1867 and at Escher's Northeast Railway Company expense he visited the International Exposition there, where Eiffel had participated in the building of the major pavilion there; cf. Stutz, 1983, p.172–180.

[147] Stutz, 1983, p.119–180.

[148] Student of Semper from 1858–1860.

[149] A synapse of the chair of architecture between 1855–1980 in: Bergier et al., 1980, p.642.

[150] In 1957 and 1958 professors replaced this with "bath house" or "atelier house"; Ronner, 1971.

Emil Roth's "study year of 1912". The prominent polytechnician and architect remarked on his years of study and on G. Gull, his design teacher of 1912: "The theme of this semester's work is titled: 'garden restaurant in the renaissance style'; [...] with this ideal project in mind he guides our steps in the process of design; other basic schemes – that might have resulted in a design of equal standard – are rejected; even further along each instance of risk-taking and new assessment is frowned upon; like a God he thinks he can form us insignificant persons and our architectonic cubes according to his image and his standards. [...] My self-confidence wavers; I lose myself in details, limit myself to designing the requested country home's small garden door and gateway, waste weeks and weeks of stubborn efforts to provide the garden gate with a metalwork design of a peacock doing cartwheels; [...] And then a private lecturer named Maillart announced his lecture on 'The form of ferro-concrete as a function of inner tension' intended for civil engineers, but which is also advantageous to myself."[151]

[151] Roth, 1971, p.106f.

Although Swiss Federal Institute of Technology's (ETH's) class of architecture experienced an upswing after Karl Moser was appointed to the school in 1915, its model of instruction – according to statements by another prominent architect, Hannes Schmidt (ETH-diploma 1918) – still remained defined by "style".[152] Yet Werner Max Moser (ETH-diploma 1921), the third important modern architect to be mentioned here, also mentions that under Karl Moser "new trends appeared characterized by a diminishing of the ossified dictate of form and by a more generous exchange between the teacher and student."[153]

[152] Schmidt, 1971, p.8–11.

[153] Moser, 1971, p.13–15.

One of the forerunners of modern methods of teaching architectonic design was Alfred Roth, who was appointed to the ETH in 1955 in the hundredth year of its history. Giedion, who lectured at the ETH between 1947 and 1958, also worked in this direction.[154] However, not until Bernhard Hoesli (1959/60) was appointed could a shift in paradigm occur, where for the first time in the history of teaching architecture a didactically novel Basic Design Course ("Grundkurs") was created under the cooperation of Heinz Ronner (construction/structure) and Hans Ess (drawing). This tradition was continued by Herbert E. Kramel until the year 1996.

[154] Humber, 1987, p.39.

Sources and References to Further Reading for Austria, Germany and Switzerland

Bandle, M., Von der Industrieschule über die Oberrealschule zum Mathematisch-Naturwissenschaftlichen Gymnasium, in: M. Bandle, B. Quadri (Ed.), Biographie einer Schule. Mathematisch-Naturwissenschaftliches Gymnasium Rämibühl. Ein Kapitel Zürcher Schulgeschichte (1832–1992), Zürich 1994, p. 10ff.

Bergier, J.-F., H. Grob, H.W. Tobler (Ed.), Festschrift zum 125jährigen Bestehen der Eidgenössischen Technischen Hochschule Zürich 1855–1980, Zürich 1980.

Bergier, J.-F., Wirtschaftsgeschichte der Schweiz. Von den Anfängen bis zur Gegenwart, Zürich 1990.

Bertrand, E., L'Enseignement Technique en Allemagne et en France, Paris 1914.

Callot, J.-P., Histoire de l'École Polytechnique, Paris 1982.

Dadda, L. (Ed.), Il Politecnico di Milano. Una Scuola nella formazione della società industriale 1863–1914, Milano 1981.

Düwell, K., Hephaistos' Schüler im 20. Jahrhundert. Zur deutschen Ingenieurausbildung und ihren Reformen seit 1899, in: T. Buddensieg, H. Rogge (Ed.), Die nützlichen Künste. Gestaltende Technik und Bildende Kunst seit der Industriellen Revolution (exhib. cat. on the occasion of the 125th anniversary of the Verein Deutscher Ingenieure VDI), Berlin 1981, p. 112–117.

École Centrale. Origines et destinées de l'École Centrale des Arts et Manufactures de Paris, Paris 1981.

Emmerson, G.S., Engineering Education: A Social History, Newton Abbot, Devon and New York 1973.

Fröhlich, M., Gottfried Semper als Entwerfer und Entwurfslehrer, unpublished dissertation at the ETH Zürich 1974 (under A.M. Vogt and B. Hoesli).

Guagnini, A., Academic qualifications and professional functions in the development of the Italian engineering schools, 1859–1914, in: R. Fox, A. Guagnini (Ed.), Education, technology and industrial performance in Europe, 1850–1939, Cambridge, Great Britain, 1993, p. 171–195.

Geist, J., Karl Friedrich Schinkel, Die Bauakademie. Eine Vergegenwärtigung, Frankfurt a/M. 1993.

Giedion, S., Mechanization Takes Command. A Contribution to Anonymous History, Oxford University Press, New York 1948.

Guggenbühl, G., Geschichte der Eidgenössischen Technischen Hochschule in Zürich (on the occasion of the ETH's 100th year anniversary), Zürich 1955.

Herrmann, W., Gottfried Semper im Exil, Basel 1978.

Herrmann, W., Gottfried Semper. Theoretischer Nachlass an der ETH Zürich (catalog and commentary), Basel 1981.

Hoesli, B., E. Gradmann, P. Hofer, A. Knoepfli, H. Ronner, A. Roth, A.M Vogt (Ed.), Gottfried Semper und die Mitte des 19. Jahrhunderts, lecture volume of the Symposium from 2 – 6 December 1974, organized by the Institute for History and Theory of Architecture at the ETH Zürich, Basel 1976.

Hotz, J., Kleine Geschichte der Universität Fridericiana Karlsruhe (Technische Hochschule), Karlsruhe 1975.

Huber, D. (Ed.), Sigfried Giedion. Wege in die Öffentlichkeit, Zürich 1987.

Jacobi, C.G.J., Über die Pariser Polytechnische Schule, in: K. Weierstrass (Ed.), C.G.J. Jacobi's Gesammelte Werke, Vol. 7, Berlin 1891.

Jelinek, C., Das ständisch-polytechnische Institut zu Prag. Programm zur fünfzigjährigen Erinnerungs-Feier an die Eröffnung des Institutes, Prag, 1856.

Klemm, F., Technik. Eine Geschichte ihrer Probleme, München 1954.

Lankheit, K., Gottfried Semper und die Weltausstellung London 1851, in: B. Hoesli et al. (Ed.), 1976, p. 23–47.

Lundgreen, P., Bildung und Wirtschaftswachstum im Industrialisierungsprozeß des 19. Jahrhunderts, Berlin 1973.
Lundgreen, P., Techniker in Preußen während der frühen Industrialisierung (Ed. Historische Kommission zu Berlin), Vol. 16, Berlin 1975.
Lundgreen, P., A. Grelon (Ed.), Ingenieure in Deutschland 1770–1990, Frankfurt a/M.1994.
Manegold, K.-H., Eine École Polytechnique in Berlin. Über die im preussischen Kultusministerium in den Jahren 1820 bis 1850 erörterten Pläne zur Gründung einer höheren mathematisch-naturwissenschaftlichen Lehranstant, in: 'Technikgeschichte' (Ed. Verein Deutscher Ingenieure VDI), Vol. 33, N° 1, Düsseldorf 1966.
Marry, C., Die langsame Feminisierung des Ingenieurberufs in Frankreich: 1945–1990, in: A. Grelon, H. Stück (Ed.), Ingenieure in Frankreich, 1747–1990, Frankfurt a/M. and New York 1994, p. 239–258.
Mathys, E., Männer der Schiene. Kurzbiographien bedeutender Eisenbahnpioniere, Bern 1955.
Maurer, B., Karl Culmann und die graphische Statik, Diepholz (Germany) 1998.
Le Moël, M., and R. Saint-Paul (Ed.), Le Conservatoire National des Arts et Métiers au coeur de Paris 1794–1994, Paris 1994.
Moser, W.M., Gedanken über die Schweizer Architektur 1916–1958, in: H. Ronner (Ed.), Die Architekturabteilung der Eidgenössischen Technischen Hochschule 1916–1956 (Arbeitsberichte der Architekturabteilung ETH, Vol. 2), Zürich 1971, p. 13–15.
Neuwirth, J., Die K.K. Technische Hochschule in Wien 1815–1915. Gedenkschrift, Wien 1915.
Niggli, P., Rector of the ETH (1930), Introduction in: Festschrift on the 75th Anniversary of the Swiss Federal Institute of Technology in Zürich, Zürich 1930.
Oechsli, W., Geschichte der Gründung des Eidg. Polytechnikums mit einer Übersicht seiner Entwicklung 1855–1905 (Festschrift on its 50th Anniversary), 2 Vols., Frauenfeld 1905.
Paschoud, M., Notice historique, in: École Polytechnique de l'Université de Lausanne. Ouvrage publié à l'occasion de son Centenaire 1853–1953, Lausanne 1953, p. 15–25.
Peters, T.F., Building the Nineteenth Century, The MIT Press, Cambridge, Mass. and London 1996.
Programme der eidgenössischen polytechnischen Schule für die Schuljahre 1872/73 resp. 1877/78, Zürich 1872 resp. 1877 [Scientific and Historical Collections at the ETH Main Library Zürich].
Reising, G., Kunst, Industrie und Gesellschaft. Gottfried Semper in England, in: B. Hoesli et al. (Ed.), 1976, p. 49–66.
Ricken, H., Der Bauingenieur. Geschichte eines Berufes, Berlin 1994.
Ronner, H. (Ed.), Arbeitsberichte der Architekturabteilung ETH, 3 Vols., Zürich 1971.
Roth, E., Studienjahre 1912, in: H. Ronner (Ed.), Die Bauschule am Eidgenössischen Polytechnikum Zürich 1855–1915 (Arbeitsberichte der Architekturabteilung ETH, Vol. 1), Zürich 1971, p. 106f.
Schirmer, W., J. Göricke, Architekten der Fridericiana. Skizzen und Entwürfe seit Friedrich Weinbrenner, in: 150 Jahre Universität Karlsruhe 1825–1975 (exhib. cat., Ed. W. Schirmer, J. Göricke), 'Fridericiana', Journal of the Universität Karlsruhe, Vol. 18 (Anniversary Volume), Karlsruhe 1975.
Schirmer, W., Lehrer – Schüler, in: id. (Ed.), Friedrich Weinbrenner 1766–1826 (exhib. cat.), Karlsruhe 1977, p. 131–134.
Schmidt, H., Schweizer Architektur der Jahre 1890–1918, in: H. Ronner (Ed.), Die Architekturabteilung der Eidgenössischen Technischen Hochschule 1916–1956 (Arbeitsberichte der Architekturabteilung ETH, Vol. 2), Zürich 1971, p. 8–11.
Schnabel, F., Die Anfänge des technischen Hochschulwesens. Festschrift anläßlich des 100jährigen Bestehens der Technischen Hochschule Fridericiana zu Karlsruhe, Karlsruhe 1925.

Schoen, J.G., Die Technischen Hochschulen und deren Organisation in Oesterreich, Leipzig 1882.
Schwarz, K. (Ed.), 100 Jahre Technische Universität Berlin 1879–1979 (exhib. cat.), Berlin 1979.
Schwarz, K. (Ed.), Von der Bauakademie zur Technischen Universität. 200 Jahre Forschung und Lehre, Berlin 1999.
'Schweizerische Bauzeitung', 100 Jahre Eidg. Technische Hochschule, Special Issue, N° 42 (15 October 1955), Zürich 1955.
'Schweizerische Hochschulzeitung', 100 Jahre Eidgenössische Technische Hochschule 1855–1955, 28th year, Special Issue, Zürich 1955.
Schweizerischer Verband der Akademikerinnen (Ed.), Das Frauenstudium an den Schweizer Hochschulen, Zürich 1928.
Semper, G., Wissenschaft, Industrie und Kunst. Vorschläge zur Anregung Nationalen Kunstgefühles bei dem Schluss der Londoner Industrieausstellung, Braunschweig 1852.
Semper, G., Der Stil in den technischen und tektonischen Künsten oder praktische Ästhetik. Ein Handbuch für Techniker, Künstler und Kunstfreunde, Vol. 1 and 2, Frankfurt a/M. 1860 resp. 1863 (New ed.: Mittenwald 1977).
Semper, G., Bericht über die Abteilung für Architektur-, Metall- und Möbeltechnik und praktisches Entwerfen; published in: First report of the Department of science and art (1854), p. 210ff.; in: H. and M. Semper (Ed.), Gottfried Semper. Kleine Schriften, Mittenwald 1979, p. 95–99.
Semper, G., Unterrichtsplan für die Abteilung für die Metall- und Möbeltechnik; published in: First report of the Department of practical art (1853), p. 372ff.; in: H. and M. Semper (Ed.), Gottfried Semper. Kleine Schriften, Mittenwald 1979, p. 100–104.
Semper, G., Über Wintergärten, in: 'Zeitschrift für praktische Baukunst' (1849); in: H. and M. Semper (Ed.), Gottfried Semper. Kleine Schriften, Mittenwald 1979, p. 484–490.
Steinman, D.B., The Builders of the Bridge. The Story of John Roebling and His Son, New York 1945.
Straub, H., Die Geschichte der Bauingenieurkunst. Ein Überblick von der Antike bis in die Neuzeit (Eds. 4th edition P. Zimmermann, N. Schnitter, H.K. Straub), Basel/Boston/Berlin 1992.
Stutz, W., Bahnhöfe der Schweiz. Von den Anfängen bis zum Ersten Weltkrieg, Zürich 1983.
Urner, K., Vom Polytechnikum zur Eidgenössischen Technischen Hochschule: Die ersten hundert Jahre 1855–1955 im Überblick, in: J.-F. Bergier et al. (Ed.), Zürich 1980.
Valdenaire, A., Friedrich Weinbrenner. Sein Leben und seine Bauten, Karlsruhe 1919.
Wickenden, W.E., A Comparative Study of Engineering Education in the United States and in Europe, in: Bulletin N° 16 of the Investigation of Engineering Education (Ed. The Society for the Promotion of Engineering Education), Lancaster, Pennsylvania, 1929.
Wingler, H.M. (Ed.), Gottfried Semper. Wissenschaft, Industrie und Kunst und andere Schriften über Architektur, Kunsthandwerk und Kunstunterricht, Neue Bauhausbücher, Mainz 1966.
Wolf, R., Das Schweizerische Polytechnikum. Historische Skizze zur Feier des 25jährigen Jubiläums im Juli 1880, Zürich 1880.
Wolzogen, A. v., Schinkel als Architekt, Maler und Kunstphilosoph (1864), in: Karl Friedrich Schinkel, Sein Wirken als Architekt, Berlin 1981.

American Lines of Tradition: From Shop Culture to School Culture

Specific efforts undertaken in America to apply the sciences to fields of technology, industry and engineering were motivated by the need to make accessible new territories and energy sources that would solve America's problems of transportation and supply in its westward settlement of the New World. Moreover, the application of scientific thought was to also increase the quality of industrial products so that the general demand for modern conveniences could be matched – and not just exclusively for East Coast urban society. However, initial scientific efforts were primarily aimed at pragmatically raising the level of professionalism among farmers, mechanics and craftsmen; efforts to establish professional technical schools were first met with some skepticism.

Nevertheless, in connection with Western frontier settlement and its respective communication requirements (such as river and canal shipping for market supplies and railways as an extensive transport network), a system of scientific education in engineering began to emerge. In this process it first augmented and then finally superseded the principle of "self-taught" or "workshop" skills and processes of immediate necessity and utility characteristic of the first generation of technical pioneers. The principle of shop culture, which is an extension of the British system of apprenticeship, was maintained for quite a while as a parallel movement to America's emerging school culture.

Due to the extraordinary demands made upon individuals and their common fate (as a historical mission) in the form of extreme territorial, climate and other natural conditions, the entrepreneurial movement imported from Europe developed along very different lines than that of the "Old World", in that the most elementary foundations of a new culture and tradition had yet to be created.

Industrial development progressed exceedingly fast and was fueled by the genuine inventions of many ingenious individuals and smaller groups who were spread across the country and were directly confronted by existential demands. These pioneers laid the foundations for "Yankee technology"[1]. The establishment of a scientific, technical and industrial system of education was completed in an equally rapid pace. These undertakings were likewise influenced by certain individuals or groups throughout the 19th century. Because America's generations of emigrants were traditionally oriented towards Europe, American educational pioneers imitated European examples that functioned well and efficiently, especially those of France. As examples they naturally had to be adapted to New World conditions.

[1] A number of the most important pioneers are mentioned in Finch, 1960, e.g. in Chapter 13 (American Civil Engineering in the 19th Century), and in Hounshell, 1984, e.g. Chapter 4 (The MacCormick Reaper Works and American Manufacturing Technology in the Nineteenth Century).

Moreover, the French and American movements of the Enlightenment existing on the eve of the French Revolution and following American independence had already established contact with each other and had laid the preliminary foundations for future relations. At this time it was considered a matter of course that science, technology as well as manufacturing and industrial development were to serve the welfare of human society and the betterment of each individual's conditions of life. Joining the concepts of human rights and democracy, the establishment and guarantee of which was to proceed by means of general education and upbringing, was actually the product of a "transatlantic spark" that took place during the era of the "Lumières".

American independence and democracy movement and the French Enlightenment. American pioneers of independence and democracy enjoyed many and diverse strong relations with the philosophers and scholars of the French Enlightenment. The Declaration of Independence of July 4th, 1776, in which colonial territories along the east coast of the American continent declared their independence from England, their Mother Country, and the Bill of Rights passed in Virginia immediately before that (June 12th, 1776), awakened the interest of those Frenchmen who were equally liberal-minded and who subsequently looked toward America with interest. This is where democracy was to emerge based on the ideas of the "Lumières"; for the first time in history, equality among men, human rights and a democratic constitution were to become established within a the "United States of America" without interference from the aristocracy. Benjamin Franklin's first formal visit to Paris on 3 December 1776 increased France's enthusiasm for America; he not only came as Ambassador of the new American government but also as a scholar. Ever since Franklin paid his first visit to Paris in 1767 he continued to maintain contact with Enlightenment scholars, including Turgot and Mirabeau. Because of his scientific work on electricity he was accepted to the Académie des Sciences in 1772. In this period meetings took place in the salons of Mme Hélvétius and as of 1786 at Mme Condorcet's, becoming important sites for the transatlantic exchange of ideas (they included, among others, Thomas Jefferson, Franklin's successor in Paris, and Thomas Paine).[2]

[2] Badinter, 1989, p.59–162.

The Beginnings of American Science

The classical British college system based on the Oxford and Cambridge model still predominated at the time Franklin and Jefferson undertook their early efforts in the 1780s to establish the new sciences, using also the French model. It was not until circles of merchants and manufactures provided the important impetus after 1820 that the applied sciences became established.[3]

[3] The documentation in this section on the USA follows W.E. Wickenden, 1929, p.58–74; Calvert, 1967, p.1–105; Emmerson, 1973, p.132–165; Donovan, 1993, p.255–276.

An example of one of the earliest New World efforts to organize the sciences is the American Philosophical Society, founded by Franklin in Philadelphia in the year 1751.[4] At the heart of this Society was its library, a place of meeting and education at the same time. Franklin also founded the College of Philadelphia at this time, which was to carry the idea of education out into the world and which laid the foundation for the later University of Pennsylvania. Not only did Franklin promote scientific culture in Philadelphia but he also

[4] This section follows Greene, 1984, p.37–106.

The University of Pennsylvania's auditorium with laboratory; the facilities for experiments and demonstration were invented by Robert Hare, who furnished the auditorium with his equipment and instruments as of 1818. (Greene, 1984, p.186)

made sure that it spread throughout all of America. When Benjamin Franklin died and Jefferson returned to America from Paris in 1790 it was the latter who not only carried on the former's scientific and cultivated spirit but also his specific initiatives.[5]

In New England it was primarily Harvard College in Boston, Massachusetts that influenced an entire region comprised of many municipal districts. Since 1761 Boston had maintained a Philosophy Club with a library, meeting hall, instrument collection, study rooms, etc. John Adams was the leading figure in organizing the intellectual culture there.[6]

Philadelphia and Boston's efforts to establish and disseminate scientific culture in America also had an impact on other regions. These scientific communities organized lectures, their members experimented in laboratories, wrote treatises, issued journals, opened libraries, museums and botanical gardens; what was important was that public educational lectures were always held for all individuals independent of their social and educational background: "Saturday night lectures" similar to efforts undertaken by the Conservatoire des Arts et Métiers in Paris or to British initiatives as well were aimed at spreading knowledge and discoveries, at stimulating scientific and methodical processes of individual activity and connecting the intellectual culture of scientific societies with the practice of daily life, of creative employment and of commercial and industrial work. Members of such scientific societies were also continuously concerned with creating further associations, such as the Linné Societies or philosophical clubs, and with creating a university or technical system of education. These efforts – parallel to shop culture – were to feed the development of education in engineering.

The Machine Shop

The historical period that defines the young American democracy up to the Civil War (1862–1865), and particularly the time between 1830 and 1860, is characterized by an emerging process of industrialization where important principles were developed that, after the Civil War, were to bring about a major industrial revolution (mass production). In this preliminary

[5] The American system of Universities was also directly influenced by the French sciences in that certain individuals emigrated to America, such as Claude Crozet, a "Polytechnicien" or Joseph Priestley, a friend of Lavoisier and a member of the circle of chemists who founded the École Polytechnique in 1794. In addition, every important treatise published by French scholars were translated into English and disseminated; cf. Greene, 1984, p.131f. or p.163f., and Calvert, 1967, p.53.

[6] In 1778 Adams paid a visit to scholars of the Enlightenment in Paris. After his return he established the Academy of Art and Sciences in 1780. This had a stimulating effect on many other circles, and as a consequence scientific institutes were founded, e.g. the Massachusetts Society for Promoting Agriculture, the Massachusetts Historical Society or the Boston Atheneum (founded in 1804), where an international exchange of ideas took place similar to Paris and Philadelphia. A comprehensive library existed with rich endowments and reading rooms where, for instance, the "Journal de l'École Polytechnique" could be consulted; the library also included a collection of paintings and sculptures.

phase mechanics and machine shops played a leading role in producing steam engines for transport and for proto-industrial production (i.e. inland shipping and later international shipping; railroads; cotton mills; mechanization of agriculture, etc.).

Workshops were in the position to not only meet the demands made by the pioneer generation but also to produce step-by-step developments in technology. Problems were solved pragmatically, theory was less important than praxis. Young mechanics were trained in the workshop – on the job. Until 1860 the shops produced everything necessary to satisfy the growing needs of an expanding and demanding American society: machine tools, locomotives, ships, textile mills, house and agricultural equipment, and so on. Not until the middle of the century did specialization occur with the beginning of precision industry.

The shops gave major impetus to industrial development and to the creation of larger growing businesses and factories. Most often a typical business would begin with a one-room workshop, then a smithy would be added followed by a foundry and so on until it gradually developed into a factory. Here only MacCormick of Chicago will be mentioned as an example of a rapidly growing factory: it produced agricultural equipment and gained particular recognition for its reaping machine exhibited at the "Great Exhibition" in London in 1851.[7] Early shops also often included a drawing studio and a business office. For each assignment construction designs were completed and stored in archives.

[7] On the MacCormick story, cf. Hounshell, 1985, p.153–187; The Great Exhibition of 1851, 1950, p.82.

At this time production was completed according to order. However, since the shops also produced technical innovations it was possible for them to offer technically improved products independently of any specific customer. A next step comprised efforts to standardize machine components and other similar parts as well as to rationalize procedures. After the Civil War a shift in paradigm occurred from commissioned to market production. Shops in this early phase were also involved in an active exchange of information with regard to non-patentable matters, which resulted in the rapid increase of general technical knowledge.

Construction office and drawing studio of the Niles Tool Works in Hamilton, Ohio around 1891. (Calvert, 1967, p.214)

Working and learning in commercial workshops entailed a process of socialization for all involved: not only did engineers work side-by-side with technicians and apprentices, but each engineer had at some point worked in a shop himself, thus becoming personally acquainted with the work, procedures, material and conditions demanded by teamwork – based on mutual respect. It was in this context that Calvert formulated the term "shop culture" to characterize the inventive, innovative, productive and social aspects of the American system of workshops. The shops' mutual exchange of information, the horizontal mobility of its personnel typical of that culture, i.e. exchange of employment from business to business, and their principle of partnership (corporations did not become established until later), etc., facilitated the creation of an engineering elite. The shop system embodied an amalgamation of intellectual, technical and social culture without sacrificing business.[8]

[8] Calvert, 1967, p.12f.

The Beginnings of American Engineering Education

The establishment of technical education for architects and engineers based on scientific principles was developed from two genuine American sources: from the efforts and institutions of locally and regionally organized scientific circles and from the experiences and

Richard Norris & Sons Locomotive Works, Philadelphia (Pa.) in the year 1855 as an example of a mechanics shop. (Calvert, 1967, p.34)

successes of shop culture. While theoretical principles and scientific and technical philosophic thought contributed to engineering education, the shops provided empirical thought and an innovative form of practice.

One of the first engineers to emerge was therefor the mechanical engineer coming out of a machine shop culture that, as has already been mentioned, was invested with an industrial quality. His first fields of activity were in machine construction and ship building, representing early mechanical and agricultural industry. Towards the end of the 1830s the railroad shops brought forth the railway engineer, who as yet still left the technical and mechanical aspects to the mechanic and instead applied himself to organization, supervision, management and logistical planning.[9] Finch, an American technological historian, argues that the construction of the Baltimore & Ohio Railroad became an actual school of engineering[10] where young engineers were trained by a process of solving many novel forms of practical assignments and where the knowledge and skills acquired there were subsequently applied elsewhere.[11] Finch considers a similar and earlier instance of teaching and learning to be typified in canal construction for inland shipping, which, with regard to the construction of the famous and at that time longest Erie Canal, he describes as "the First American Engineering School".[12]

[9] With respect to privately organized competitive companies – this particularly included solving problems of space–time economy: travel time and schedules, transport weights and costs, geographic distances and person statistics, etc.

[10] "The First American Railroad Engineering School".

[11] Finch, 1960, p.268f.

[12] Finch, 1960, p.262–264.

The stimulus to create a systematic and high-quality education for engineers based on scientific principles was provided by military maritime shipping. The necessity of building a navy that was guaranteed to be seaworthy and capable of sustained deployment without developing functional flaws created the demand for fundamentally better trained engineers who were also to enjoy higher social status. In this respect, when the first schools were founded the French example of the École Polytechnique and its related specialized schools were drawn upon as models.[13] The general sciences, i.e. mathematics and physics, predominated over all other specialized fields of profession, even over mechanical engineering. Under Robert H. Thurston, one of the Naval Academy's leading professors who also considered Prussian schools exemplary, scientific studies were combined with practical courses. Calvert emphasizes, however, that until about 1850 shop culture still remained the major source for recruiting young aspiring engineers.[14]

Besides the fields mentioned above, further impetus for developing a systematic scientific and technical engineering education was provided by bridge building. Particularly here pioneers reached the limits of their skill in endeavoring to match ever higher demands that exceeded their practical experiences gained from within shop culture. It is therefore not coincidental that the most daring bridge builders of the pioneering age, namely still before the Civil War, were either graduates of already existing schools of engineering, such as Squire Whipple and Jacob Hayes Linville who both studied at the Rensselaer Polytechic Institute (Union College)[15], or were immigrants from Europe such as Albert Fink and Johann August Roebling from Germany.[16]

The First Schools of Engineering

The first school of applied sciences to be established in America was the *Military Academy in West Point (New York)*, initiated in 1798 by President Adams, formally inaugurated in 1802 and instituted as a school of engineering as of 1818. Colonel Sylvanus Thayer directed the school as of 1817. He was a graduate of Dartmouth College (1807) and also of West Point

[13] The Military Academy in West Point, New York (founded 1802 and school of engineering as of 1818) and the Naval Academy in Annapolis, Maryland (inauguration 1845).

[14] Calvert, 1967, p.20–23.

[15] Linville built Schuylkill River Bridge in 1861.

[16] Roebling studied at the Berlin Bauakademie (not at the "Royal Polytechnic Institute" as incorrectly stated in Steinman's biography) as of 1823 under Eytelwein (construction statics, mechanical and hydraulic engineering) and Dietleyn (bridge building), in 1831 he emigrated to America, built the suspension bridge over Niagara between 1852–1855 (together with Charles Ellet), and in 1869–1883 he built the Brooklyn Bridge in New York, which was completed by his son Washington August Roebling, graduate of the Rensselaer Polytechnic Institute in 1857; cf. Finch, 1960, p.280–283; on John A. Roebling, cf. Steinman,

(1808). Thayer subsequently undertook an assessment of European military schools and of France's military engineering system and began to reorganize his school according to the example provided by the Parisian École Polytechnique. The school's teaching program focused on the natural sciences (chemistry and physics), mathematics and a broad field of engineering applications that correlated to the French model's Cours d'application. In 1817 one hundred and fifty cadets studied at West Point. Of the 121 graduates to already come out of the school between 1802 and 1827 fifty-seven chose a military career. In the beginning of Thayer's office term 15 professors were employed, including Claude Crozet[17], who was appointed by Thayer himself.

Shortly after West Point and at the same time as Karlsruhe (in 1825) the *Rensselaer School*, founded by Stephen Van Rensselaer[18] in 1823 as part of Union College in Troy (New York) and under the directorship of Amos Eaton, introduced continuing education courses for advanced students, thus providing the school with a further dimension in engineering. In the instruction program Rensselaer states that the sciences to be taught at the school (experimental chemistry, philosophy and natural history) were to serve the general welfare and that graduates of the school were to contribute to this by disseminating and teaching their application in all fields of agriculture, domestic sciences, commerce and trade.[19]

Rensselaer's programmatic concepts entailed a combination of subjects that five years later were to be realized by the Parisian École Centrale des Arts et Manufactures; however, industrial development and raising the national scientific and technological standards were not being aspired to here, but rather local elementary needs were being taken into consideration. Besides the agricultural sciences, school focus was laid on land surveying as of 1827, which, within the context of the Land Ordinance Act of 1785 and in connection with regional land ordinance laws, became an increasingly important task of engineering all over the country.[20] In 1828 the term civil engineering was used for the first time in the school program to describe the sum total of all scientifically based areas of

1945; on Washington A. Roebling, cf. Weigold, 1984. Charles Ellet (1810–1862) studied at the École des Ponts et Chaussées, returned to his home in 1832 and became one of the leading bridge building engineers of America; on Ellet, cf. Emmerson, 1973, p.139 and Brown, 1993.

[17] Crozet was a graduate of the École Polytechnique, he was to teach engineering at West Point, however he first introduced "Géometrie descriptive" according to his teacher Monge and in 1821 wrote his own textbook in English titled "Treatise on Descriptive Geometry". After teaching he became senior engineer of the State of Virginia and built "America's best mountain roads" there; cf. Greene, 1984, p.131f.; Wickenden, 1929, p.62; Emmerson, 1973, p.140f.

[18] Stephen Van Rensseler was an offspring of Dutch immigrants, major land-owner, financier and politician from Albany, New York; he received his scientific qualification from Harvard (graduated 1782).

[19] Wickenden, 1929, p.61.

[20] Syrett, 1960, p.94 (on the Land Ordinance Act), and p.97ff. (on the Northwest Ordinance).

application in engineering education. From this an integral program of instruction developed in 1835 titled "engineering and technology" for which the school awarded a "Civil Engineer" diploma for the first time.

Rensselaer engaged Amos Eaton to be the driving force behind the school's program and organization. Eaton, as a graduate of Williams College (in arts, 1799) and as a student of Silliman at Yale, had acquired a great deal of scientific and practical knowledge in a wide range of areas.[21] He established a post-graduate study with focus on the natural sciences as an extension of college and with the primary aim to contribute to agricultural industrialization. This, at least, was his first concern.[22] Within twelve years as its director Eaton raised the school to a high professional niveau and created a prototype of the modern American post-graduate school; it was *the first scientific school of engineering in the entire English speaking world* and it exerted a great deal of attraction on students and teachers alike.[23]

Amos Eaton's student and successor, B. Franklin Greene, completely reorganized the school in 1847. He likewise undertook an analyses of various leading European models of instruction and subsequently endeavored to establish a comprehensive polytechnical institution of architecture and engineering.[24] The preceding Rensselaer School was to continue to provide preparatory scientific education according to the model provided by French Collèges.[25]

The evident pedagogical dimension introduced by Greene's in the education of architects and engineers is to be considered as the continuation of the Enlightenment tradition which saw pedagogical culture as part of raising the general and individual welfare of all citizens. Greene did not apply the term *Rensselaer Polytechnic Institute* until the year 1849. The name expresses reverence for the school's polytechnical roots on the one hand and on the other hand marks a turning point away from the primate of agriculture established by the first entrepreneurial years. The program was limited to two courses of study, civil engineering and topographical engineering. For this Greene drew upon the meanwhile newly established (1828/29) École Centrale in Paris.[26] Thanks to these reorganizing efforts

[21] Wickenden, 1929, p.61.

[22] According to Emmerson, the Prussian system of technical education provided Eaton with an example. However, the curriculum later changed as the importance of agriculture was replaced by industry. Eaton was one of the first to commit himself to promoting women's study of the sciences in America; Emmerson, 1973, p.144–146.

[23] Emmerson, 1973, p.144–146.

[24] Wickenden, 1929, p.63.

[25] Wickenden emphasizes the importance of Greene's plans: "Without doubt Greene was the first man in America to submit the problems of education for the technical professions to thorough investigation and analysis." Wickenden, 1929, p.63.

[26] Wickenden, 1929, p.63.

the Rensselaer Polytechnic Institute was able to reinforce its status as an industrial vanguard.

In comparing the curricula of the Rensselaer Polytechnic Institute (1849/50), the École Centrale in Paris (1850), MIT (1865) and the University of Illinois (1867) it becomes apparent that the basic sciences and their technical and industrial fields of application were being taught parallel to humanistic studies, which has typified the system of education for architects and engineers in America up to the present. Humanistic studies particularly included English and a foreign language, partly history and philosophy, though the spectrum varied from school to school.[27]

[27] Wickenden, 1929, p.64; Emmerson, 1973, p.154–155. A further pioneering school is represented by the "American Literary, Scientific and Military Academy" founded in Norwich, Vermont in 1820 by Alden Partridge, former professor of West Point (in 1834 it received university status); cf. Emmerson, 1973, p.141ff. Also, the "Gardiner Lyceum" founded in Maine in 1821 by R.H. Gardiner is to be considered as one of the most exceptional early schools of scientific and technical edcuation; Benjamin Hale taught here (due to a lack of funds the school already closed in 1832); cf. Wickenden, 1929, p.61.

Engineering Studies at the Universities

Greene's efforts acted as a stimulus to many other institutions to pursue this direction for architectural and engineering education. In this regard departments of civil engineering were instituted at various universities. In 1851, for instance, the *University of Dartmouth (New Hampshire)* established the Chandler Scientific School. Benjamin Hale, who had previously taught at the "Gardiner Lyceum", played an important role in setting up this school. In 1852 the *University of Michigan* followed with a complete course of studies for civil engineering, which included also courses on the theory of construction, of architecture, etc.[28] Its founder De Volson Wood was previously a teacher at the Rensselaer School.[29]

[28] Emmerson, 1973, p.151f.

[29] William Le Baron Jenney was employed here between 1876 and 1880 as professor of architecture.

In the year 1847 the universities of Harvard and Yale introduced courses on applied sciences. One of the initiators at Harvard was Prof. Eben N. Horsford, who in turn was a graduate of the Rensselaer Polytechnic Institute. Thanks to an endowment presented by Abbot Lawrence the engineering institute *Lawrence Scientific School at Harvard* became established.[30] Lawrence sought to react to the challenges posed by industrialization by reinforcing the three pillars of scientific education, namely engineering, mining and metallurgy, and mechanical engineering. In his opinion they all originated from the same basic scientific principles and only in their application did they become different.[31] Lawrence thus followed the French tradition based on the Parisian École Centrale's creed "la science est une"

[30] This is where William Le Baron Jenney studied in 1851 before going on to the Parisian École Centrale des Arts et Manufactures.

[31] Emmerson, 1973, p.148f.

("a sole science"). Nevertheless, this institute was not efficient enough – after 45 years only 155 engineers had received their diploma from this school – resulting in the foundation of the Massachusetts Institute of Technology (MIT) in the year 1860.

At the behest of Benjamin Silliman *Yale* began to offer courses on applied chemistry, and not until 1852 and upon the initiative of John P. Norton[32] was this followed by mathematics and architectural engineering (including courses on, e.g. civil engineering, principles of architecture, architectural drawing with shading and tinting, etc.)[33]. In 1856 the school instituted a complete course of three-year studies on civil engineering and also, for the first time at an American university, mechanical engineering, which existed only on paper for a few years until a sponsor could be found in the person of Joseph E. Sheffield. From this the *Sheffield Scientific School* was established.[34]

[32] Emmerson mentions William A. Norton (p.149).

[33] Documentation of curriculum in Emmerson, 1973, p.149f.

[34] Wickenden, 1929, p.64–66.

Morrill Land Grant Act of 1862: Turning Point for School Culture

Until 1860 the pioneering spirit of far-sighted individuals such as Franklin, Jefferson, Thayer, Van Rensselaer, Lawrence and Sheffield dominated the scientifically based technical and industrial education of architects and engineers. However, their efforts were but first steps in producing an entire system of education in engineering based on Enlightenment concepts.

A decisive turning point in the development of a public system of higher learning did not occur until the "Morrill Land Grant Act" was passed in 1862.[35] In the Morrill Act, named after its initiator Senator Justin Morrill (Vermont), the Federal Government granted land tracts or property rights[36] to individual states in order to finance (through agriculture or mining) new schools. Existing schools mostly expanded their course offer and established departments of mechanics, industrial training, dynamic or mechanical engineering. The Morrill Act launched an entire school movement and spurred the establishment of many new technical institutes, as part of colleges as well universities.[37]

While between the year 1830 and 1860 the number of university institutes and colleges increased from

[35] After first efforts were attacked by conservatives in 1857 second efforts proved more successful five years later, in the middle of the Civil War; in 1862 congress was dominated by the Northern states.

[36] 30,000 acres were granted for each senator and congressman; cf. Syrett, 1960, p.279.

[37] Calvert, 1967, p.47f.

100 to 500[38], the success posed by the Morrill Land Grant Act was not immediately visible.[39] The first school to be established as a consequence did not open until around 1870 after the Civil War with the *Stevens Institute of Technology* in Hoboken (New Jersey) founded by Edwin A. Stevens. Henry Morton, physicist and secretary of the Franklin Institute held the office of first president and professor of the school. He engaged Robert H. Thurston to establish courses and a curriculum in mechanical engineering as of 1871 as a model of pure professional training. Thurston remained active until 1885. The Stevens Institute initiated other similar schools, such as *Purdue University* in Indiana – one of the first institutes to become augmented by a "land-grant college", or *St. Johns College* in Annapolis (Maryland) and the *University of Illinois,* which became equipped with the most modern laboratory facilities for practical courses as of 1883.

[38] The population increased in the same amount of time from 12 to 31 million; Emmerson, 1973, p.146 and 138.

[39] The ensuing follows Calvert, 1967, p.49–51.

As an example of the transition from the traditional shop culture to school culture is *Sibley College of Engineering at Cornell University* in Ithaca (New York). Sibley College was set up as a consequence of the Morrill Act by the State of New York. Already in 1865 John L. Morris instituted a mechanical arts department at Cornell University. Its specific goal was to make available an equally professional educational situation in comparison to practice in "engineering shops" at the college itself. John E. Sweet was engaged to direct these workshops as of 1873. His achievements included not only the establishment of functional workshops but also the integration of practice courses in school; the shops therefore became an integral part of intellectual culture.[40]

[40] Already in 1879 Sweet, a typical advocate of shop culture, left the school because he was denied the acquisition of further equipment to set up the workshops.

Around 1880 an increased trend towards the "pure sciences" and to the primate of theory became evident in the U.S.. As a result, the shops and the schools lost their importance. As of 1885 Robert H. Thurston, who up until then had taught at the Stevens Institute and was an advocate of school culture, was employed at Sibley College. Through his efforts education in theory became predominant. Cornell did not only become one of the most leading institutes of the American education movement with regard to a theoretically based education in engineering but also many of this school's

Silbley College of Engineering at Cornell University in Ithaca (New York) around 1885, entire facility and machine shop. (Calvert 1967, p.181)

graduates in turn established their own schools or taught at similar institutions.[41]

The University Movement after 1862
The driving forces behind the development of institutionalized education in engineering included besides Cornell University also essentially the Massachusetts Institute of Technology and the School of Mines at Columbia.

The founder of the *Massachusetts Institute of Technology* was William Barton Rogers, geologist and pedagogue. Due to the support provided by influential Boston circles and by the State of Masssachusetts financing, MIT was already able to institute a school of engineering as of 1861. Rogers' "Objects and Plan of

[41] Wickenden, 1929, p.68f., and Calvert, 1967, p.87–105.

an Institute of Technology" of 1861[42] provided for a school of industrial science and a society of arts as well as a technical museum. While the term "industrial science" originated from the Parisian École Centrale, the latter makes reference to earlier projects, e.g. to those of Condorcet.[43]

[42] Calvert, 1967, p.44.

[43] Particularly Condorcet's "Projet sur l'instruction publique".

The three institutions – the school, the scientific society and the museum – were to bring about progress in the applied sciences in the fields of technology, agriculture, industrial production and trade. The society of arts was responsible for disseminating scientific knowledge amongst the public. While the society was able to begin its work immediately (1862), the school of industrial science did not open its doors until 1865; the museum, however, was not realized for quite some time. From the very beginning MIT was the institute in the education of America's architects and engineers to offer the most comprehensive program of courses in the applied sciences, including architecture, shipbuilding and other fields of engineering.[44] Wickenden compares it to the major German technical universities of the time. An innovation of teaching and methodology is represented by the introduction of laboratory instruction with individual work stations for the physical sciences.

[44] Wickenden, 1929, p.67.

The iron and coal industry of the Middle West were aslo in great need of scientifically trained cadres as of 1850. These were initially recruited from the circle of American graduates of German mining schools, especially from that of the Freiberg Bergakademie founded in 1765, and later from the French École des Mines opened in the middle of the 1780s. The metallurgy engineer Thomas Egleston, who after studying at Yale graduated from the École des Mines, established after his return in the year 1864 the first American school of mining, the *School of Mines at Columbia College* (today Columbia University in the City of New York).[45]

[45] This college had already existed since 1754 (as King's College).

The establishment of this mining school was preceded by many efforts to make available basic courses in mathematics and natural sciences until gradually leading engineers made the connection to engineering praxis, especially with respect to canal building.[46] Egleston employed other, equally European educated

[46] For an in depth account, see: Finch, 1954, chap.1 and 2.

engineers as professors such as Chandler, who studied at Harvard and Göttingen in Germany, and Charles A. Joy, graduate of Union College (1844) who also studied at Göttingen (1852), and more.[47] This school also introduced a parallel course of study in the humanities. This was followed by many other new schools.

[47] Emmerson, 1973, p.152.

Architects of Modern America and their Education

The professionalism of American engineers and architects was formed by an interplay of shop culture, scientifically based engineering education and professional organizations.[48] Together, like everywhere else, they outlined the career prospects open to the individual and completed his socialization within a specific professional field. While the educational environment defined the standards of knowledge and skills, the professional organizations watched over the titles, norms and formal examinations for work qualifications. What is interesting here is that Cyrus MacCormick, for example, who began his career as a blacksmith and turned his company into one of the largest producers of agricultural machines within forty years, was able to work on an equally high level and become just as recognized as someone like William Le Baron Jenney, who attended one of the best French schools, taught at the University of Michigan and founded the “Chicago School of Architecture”.

[48] The American Society of Civil Engineers was founded in 1852, four years after its respective French society was formed.

Like other sectors of industry, the development of building technology in the 19th century in America was able to draw from the two sources of “Yankee technology”: empirical and pragmatic shop culture and the scientific and technical institutes of higher learning. While construction systems and material technology were perfected in the thrust provided by canal construction, railway engineering (particularly bridge building) and the “iron age” of steel skeleton construction and iron facades, modern architecture additionally profited from a standardized system of timber framework construction (“balloon or platform frame”), manifested, for instance by the “skyscraper technology” emerging in the aftermath of the Chicago Fire of 1871, and it also profited in general from the technique of mass production.

Three ways were open to the American architect of the time to acquire professionality: he could become trained in an studio of a renowned architect, he could attend an engineering school or the École des Beaux-Arts in Paris. Often a combination of these three possibilities typified an architectural career. It must also be noted that Thomas Jefferson, who maintained close ties to the French Enlightenment, in building the Virginia State Capitol in Richmond he established America's pronounced *European* (1785–1790) *orientation* in architecture, which later developed into a sustained tradition.[49] Many architects subsequently endeavored to not only incorporate in their architectural design stylistic references (Gothic, Renaissance, Baroque, etc.) but also industrial iron technology. After the Civil War and especially as a result of Chicago's fire America's "industrial course" became reinforced, while the École des Beaux-Arts' influence in architectural education increased concurrently. This trend reached its apex in the World's Columbian Fair of 1893 in Chicago.

[49] Pevsner, 1976, p.36f.

Early Iron Constructions. The German-American architect *Benjamin Henry Latrobe* was the first one to introduce iron elements in building structures in America. While in England he had worked with the British architect Cockerell and also under Smeaton, one of the leading engineers of the British industrial era. From this his knowledge of British iron technology in building was derived. When he came to America in 1793 with Jefferson's support he initially worked in Richmond (Virginia) and then in Philadelphia in the neo-classical style. His first modern work that included iron structural elements was Christ Church in Washington D.C. (1808).[50] *William Strickland*, Latrobe's student, was the second architect to further use iron elements in architecture (Chestnut Street Theater in Philadelphia, 1820–1822 and the United States Naval Asylum in Philadelphia, 1826–1830). He was later active as an engineer in canal and railroad construction.[51] Their work paved the way for the continued and more expanded use of industrial structural systems, particularly for the use of iron structural techniques in building construction as exemplified in the architectural activity of New York, Chicago and other cities.

[50] Condit, 1982, p.79; Pevsner, 1976, p.37f; Pevsner, 1992, p.371.

[51] Condit, 1982, p.79f.; Pevsner, 1992, p.610.

James Bogardus was a watchmaker, mechanic and inventor. He was trained in cast-iron techniques by Daniel Badger. In 1848 he opened a foundry in New York and produced iron supports, girders and facades. In 1849 he submitted a patent for buildings completely made of cast-iron. A number of buildings of this type were realized by him, such as Harper's Building (architect: John B. Corlies, 1854), and in 1856 he published a manual on: "Cast-Iron Buildings. Their Construction and Advantages". Only the structure and material of his system is modern, the "style" these buildings and facades exhibited still typified the historical eclecticism of the time. Nevertheless, the design provided for the iron facades to be placed in front of the buildings' supporting structure, thus representing an early form of "curtain wall". Bogardus

Wanamaker Department Store in New York (architect: John Kellum, 1859–1868); iron skeleton structure after the 1956 fire. (Condit, 1982, fig. 26)

opened his firm on the same street as *Daniel Badger's*, who maintained a large and expanding business in the production of iron buildings and components of all kinds. His most important building is probably the Haughwout Building in New York of 1857 (architect: John P. Gaynor), in which Elisha Graves Otis installed his first person lift with an automatic break system.[52] The largest iron skeleton building of that time was the Wanamaker Department Store in New York commissioned by one of the pioneers of the American department store, Alex T. Stewart, after plans by the architect *John Kellum* and carried out by the firm Cornell Iron Works of New York (1859–1868).[53]

Around this time and directly preceding the Civil War the monumental work of building the Capitol dome in Washington D.C. began under the architect *Thomas U. Walter* in cooperation with the engineer Montgomery C. Meigs and his assistant August Schönborn.[54] The dome's support system is a complete iron structure, and in the interest of fire security even the roof tiles, window frames and other elements are made of cast-iron. Walter was born in Philadelphia of German descent, he studied and worked under Strickland; as of 1830 he set up his own office and also carried out architectural engineering commissions.[55]

Railroad Stations. Like in Europe, railway construction in America not only represented a field of experimentation for novel systems, techniques and materials of construction but also provided a source of recruitment for inventive architects and engineers and challenged them in their cooperation. By the example of a number of railroad stations originating from after the Civil War it can be demonstrated how the industrial method of construction gradually conquered architecture, becoming a predominant mark of modern railroad buildings. Cleveland Union Depot shared by the Lake Shore and Michigan Southern and the Cleveland and Pittsburgh companies (New York) is the first hall to be supported by an iron structural system in America (engineer: B.F. Morse, 1865/66). The roof construction is supported by side walls. Gradually the use of three-hinged arches, then of "Polonceau Trusses", etc.[56] followed. The most comprehensive construction sites in America at that time included Pennsylvania Station in

[52] Besides the iron businesses of Bogardus and Badger a further firm is to be considered as a pioneer in the development of rolled iron in building construction: the Trenton Iron Works founded by Peter Cooper and Abram Hewitt, who also erected buildings for Harper and Cooper Union (today one of the leading architectural schools in America). On the Iron Works' background in New York, cf. Condit, 1982, p.81–84; Pevsner, 1976, p.216f.; Pevsner, 1992, p.93.

[53] This department store comprised approx. 36,000 sq. meters. In 1956 it burned down. Condit, 1982, p.85; Pevsner, 1976, p.270.

[54] The design was based on St. Isaacs Cathedral in Petersburg by A. Ricard de Monferrand, 1842; the design for the Capitol dome in Washington was drawn up in 1856, construction began in 1864.

[55] Condit, 1982, p.90f.. Pevsner, 1992, p.678f.

[56] Thus, St. Louis Union Station was able to span 200 meters at a height of 75 meters in the middle; architects: Theodore C. Link and Edward D. Cameron, engineer: George H. Pegram, 1891–1894.

New York, where concrete structures were used in the multi-storied basement. It was built in 1903–1910 by the renowned architectural firm McKim, Mead & White (engineer: Charles W. Raymond).[57] This firm played a leading role in the education of a very influential generation of architects.

[57] Condit, 1982, p.132–138 and 179.

The New York Skyscraper. The first "skyscraper" in New York was erected between 1868 and 1870 as the Equitable Life Assurance Company office building. Arthur Gilman and George H. Kendall first designed a masoned building structure, however before construction began the young architect and engineer *George B. Post* was engaged to work over the designs. Post introduced a skeleton and roof structure similar to the ones used in the British cotton mills comprised of steel girders and filled in by arcuated masonry. With it he was able to reduce the building's weight and cost. This approx. 43–meter-high building marks the beginning of Post's career. It is followed by the 76–meter-high Western Union Building (1873–1875) and the Produce Exchange Building (1881–1884). For this Post designed the novelty of a purely iron skeleton supporting structure in which the facade is detached and thus liberated of its supporting function. The only functions remaining to it was in providing external fireproofing and stylistic appearance. The building remained in use for the next 75 years. In 1891/92 Post completed the 15–storied Havemeyer Building. Although the skeleton support structure stood independently and was designed to include the exterior wall, Post did not refrain from mantling it with masonry up to a total height of 60 meters. This method of construction typical of New York differed here from that of the Chicago School where the iron skeleton was only covered by cast terracotta elements that fireproofed the exterior and especially enabled the facade to be glassed on a grand scale.[58]

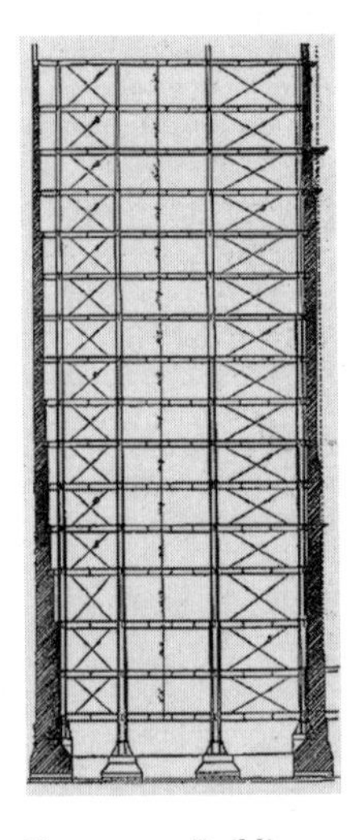

Havemeyer Building New York (1891/92) by George B. Post, architect: the iron skeleton support system stands independently, and is completely encompassed by masonry up to a height of 60 meters. (Condit, 1982, fig. 39)

[58] Condit, 1982, p.115ff. and 119.

George B. Post was trained in architecture under Richard M. Hunt, who established a studio in New York in 1857 and educated an entire generation of young architects, for instance, Henry Van Brunt, William R. Ware, Frank Furness and others.[59] Like most of Hunt's students, Post was also one of the influential architects to participate in the 1893 World's Columbi-

[59] Draper, 1977, p.120.

an Fair in Chicago.[60] *Richard Morris Hunt* was also at the forefront of skyscraper building. At the same time Post's Western Union Building was being built he erected the Tribune Tower, which was approximately 10 meters higher than Post's.[61] Hunt was educated by Hector Lefuel (cooperated on the Louvre) under the influence of the École des Beaux-Arts. He brought the neo-renaissance style to America (1855). As an institution his studio was also organized according to Paris examples.[62] He was a founding member of the American Institute of Architecture.[63] *William R. Ware*, graduate of Harvard University and a further student of Hunt, founded America's first school of architecture at the Massachusetts Institute of Technology in 1865 according to the École des Beaux-Arts model. The "School of Industrial Science" established by William B. Rogers in 1861 at MIT was thus given a more classical orientation, turning it away from education in engineering. Ware maintained an office with Henry Van Brunt, also a graduate from Harvard and student of Hunt, from 1863–1881.[64]

The examples provided by Post, Ware, Van Brunt and others demonstrate the influence of Hunt's studio and the "Manière Beaux-Arts" on the development of American architecture in the last third of the 19th century. The trend exhibited by the "Chicago School" stands in contrast to it.

The "Chicago School of Architecture"

It is probably characteristic of skyscraper technology and of the entire development of extensive iron structures that they did not become possible until after the Civil War. They emerged at a point in time in which mass production was introduced based on standardization, technical perfection and a system of interchangeable parts resulting in the functional determination of form.[65] As already elucidated, Jenney paved the way for the "Chicago School of Architecture" by a number of important buildings, by educating students in his studio such as Holabird and Roche, by influencing a whole group of leading Chicago architects (Burnham, Root, Sullivan, and others) and by inventions that were later taken up and perfected by other architects such as Mies van der Rohe (Lake Shore

[60] Post built the Manufactures and Liberal Arts Building that comprised 570 x 260 meters, representing the world's largest building of that time; Pevsner, 1976, p.250.

[61] Hunt built the Administration Building at the Columbian Fair in 1893.

[62] Hunt was one of the first Americans to study at the École des Beaux-Arts (1845–1853).

[63] Condit, 1982, p.116; Draper, 1977, p.214f.; Pevsner, 1992, p.284f.

[64] Van Brunt built the Electricity Building (together with Howe) at the World's Columbian Fair of 1893, which in its interior was equipped with a modern industrial steel frame construction which corresponded to the standard exhibited by the Paris Halles centrales built in 1853; however, its exterior was provided with an historicized facade; Pevsner, 1992, p.679f.

[65] Jenney wrote a basic article on this theme in 1890 – "An Age of Steel and Clay", which has already been treated in a previous chapter. On the American development cf. Giedion, 1941, part V.

Sullivan's rationality and décor: Carson Pirie Scott Store (1899, 1903/04). (Siry, 1988, p.2)

Drive Apartments, Chicago 1948–1951) or Skidmore, Owings & Merrill (Lever House, New York 1952). This modern building tradition is still effective today.[66] [color plate 11]

However, what Chicago's commercial office buildings lack is a spatial quality despite the existence of pronounced window segments (bay windows) and the large openings with side window sashes (Chicago windows), which both rather address issues of daylight comfort. This important architectonic criteria was stressed by Frank Lloyd Wright and later, in reference to him, by Colin Rowe in the Chicago School's architectural critique.[67] Furthermore Wright did not join the "Beaux-Arts Movement" and refused to accept his Patron Daniel Burnham's offer to study for four years at the École des Beaux-Arts at his expense.[68]

A further and different type of critique was provided by Sullivan, though having studied in 1874 at the

[66] It is scientifically oriented (in that architectonic problems are worked out by engineers, e.g. foundation statics, support structural systems, glazing technology, climate and comfort, conditioning facade, etc.) and it is industrial (construction site management, standardized building production and building process, "time is money", etc.). Cf. among others Freitag, 1904.

[67] Rowe, 1970.

[68] Draper, 1977, p.216.

18 South Michigan Ave. (Gage Group, Louis Sullivan building at right; 1898/99). (Condit, 1964, fig. 80)

École des Beaux-Arts in Paris he nevertheless condemned the excesses of this sort at the Chicago Fair of 1893. His exemplary buildings show that Sullivan developed his own unique architectonic style: the superimposition of functional rationality (support structural order, lines of reference in facades) and formal ornamentation (non-supporting filling components and curtain wall, storied bases with entrances).[69]

On the Influence of the École des Beaux-Arts on the Education of American Architects

The influence of the École des Beaux-Arts on education in architecture and on America's building culture did not come into effect until after the Civil War. It was brought about on the one hand by a few American architects who resided in Paris, such as Richard M. Hunt, who was the first to study at the school there (1845–1853), Henry H. Richardson (1859–1862), Charles F. McKim (1867–1870), Louis Sullivan (1874), as well as others.[70] On the other hand those institutions modeled on this Paris school reinforced the in-

[69] A correlation to Sempers theory of tectonics (active and passive structural elements) is unmistakable; cf. Semper, 1977, Vol. 2 (on "Metallotechnik"), p.502.

[70] The kind of education offered at the Parisian École des Beaux–Arts at the end of the 19th century is treated in detail by the example of John Galen Howard (a further student of Hunt) in: Draper, 1977, p.222–235.

fluence of the classical approach to architecture and to architectural education in the U.S.. This included the Society of Beaux-Arts Architects who awarded a “Paris Prize” as of 1894, and the American Academy in Rome with its American “Rome Prize” established in the same year by the promoters of the Chicago Fair, as well as the later Beaux-Arts Institute of Design (BAID) in New York, founded in 1916. A growing number of architecture schools were founded by students coming from established studios, e.g. MIT’ school founded by William Ware or the one established at Berkley, California by John G. Howard (both were educated at Hunt’s studio). In addition, a number of architecture schools in America engaged graduates of the École des Beaux-Arts as teachers in the 1880s, for example for MIT, Pennsylvania and Harvard.[71] The goal was to launch a “Beaux-Arts Movement” that was to delineate a national style. A further source of important influence was the studios themselves, which were managed as educational institutes according to the French tradition.[72]

[71] Roth, 1979, Chap. 5.

[72] Draper, 1977, p.216; Woods, 1999.

Reliance Building by D.H. Burnham & Company (with Charles B. Atwood), 1894/95; the industrial method of building helped to save in building material, weight, construction time and costs: the ten upper floors were assembled in 14 days. (Freitag, 1904, p.71)

The World's Columbian Exposition of 1893 in Chicago marks the high-point in which an "architectural style" was presented to the public – before the very gates of a city that brought forth the "Chicago School of Architecture" and directly next to the modern industrial pavilions built for the fair, such as the pavilions of mechanical engineering, electricity, Jenney's Horticultural Building, etc. – which made evident the increasing influence exerted by the École des Beaux-Arts on the education and activity of American architects.[73] Whether this entailed not only the "triumph of Beaux-Arts Classicism" as emphasized by Pevsner[74] but also the end of the "Chicago School of Architecture" remains questionable, especially if, for example, one were to take into consideration Holabird's and Roche's comprehensive work or Burnham and Atwood's Reliance Building (1894/95) – both of them took part in the Chicago Fair of 1893! The Reliance Building skyscraper can be considered the most light, elegant and modern building of Chicago's Loop section, a building that according to Condit, the Chicago School chronicler, was only outdone by Mies van der Rohes's Berlin "Glas-Hochhaus" project of 1919.[75]

[73] Cf. the exhibition catalog by Bancroft, 1894.

[74] Pevsner, 1976, p.250.

[75] Condit, 1964, p.110.

Conclusion

The "Chicago School of Architecture" is on the one hand a product of "engineering education"; consider, for instance, its founder Jenney. His most important student and colleague, Holabird, was also educated in engineering, namely at the United States Military Academy in West Point (1873–1875); in 1875 he was engaged by the Jenney firm as an engineer. In contrast, Roche, who worked with Jenney for almost ten years and opened his own firm with Holabird in 1881, was educated on the job. This group worked well into the 20th century and greatly influenced skyscraper technology – from Mies van der Rohe, SOM, etc. – up to the present.[76] A further protagonist of the "Chicago School" is John W. Root (the firm of Burnham & Root), graduate of the Engineering School of New York University.[77] He studied at this school from 1866 and received a Bachelor of Science degree (in civil engineering) in 1869, according to Condit the best education available in architecture in the 19th century.[78] Root's

[76] Condit, 1964, p.116ff.

[77] Boyle, 1977, p.315.

[78] Condit, 1964, p.44ff.

partner Burnham was in turn educated by studio practice under Jenney.[79]

[79] Burnham tried to do study at Harvard and Yale after receiving a college degree, however he failed to pass the entrance exams, a matter which troubled him throughout his life as he felt he lacked intellectual education; in 1868 he entered the Jenney architectural firm, staying there but a short time. Condit, 1964, p.43–69.

The Chicago School was therefore nurtured by two cultures: by an industrially oriented studio and the tradition of engineering schools on the one hand and by the master studio in the Beaux-Arts tradition on the other. Both types of studios were based on an independently and originally American development of shop culture, thus on a unique concept of learning and activity that served the immediate and vital needs of its pioneering generation.

Transatlantic Interaction Europe – America

The pioneering movement of American scientific and technical industrial schools of engineering was oriented on European models and in particular on French "original models": the École Polytechnique and École Centrale des Arts et Manufactures.

While the École Polytechnique remained due to its close ties to the state politically dependent on the French government and the École Centrale was dependent upon the government for its financial support, the respective schools that developed in the U.S. were organized through personal and private initiatives and were thus financed accordingly; therefore, the Morrill Act (1862) allowed the same type of government dependency to be avoided and relativized the government's influence on education.

After 1880 the Beaux-Arts tradition came to have ever greater influence on the activity of American architects and of many schools. Since then two fundamentally different traditions have emerged. Both were made effective by studios, offices, companies, schools and professional institutes. The Beaux-Arts movement was oriented toward French classicism, and the modern industrial movement had also evolved from European origins while rapidly adapting itself to American conditions of production, procedures and markets, thus paving the way for future building culture – in Europe as well.[80]

[80] Banham, 1986, p.215–228.

The first evidence of this to appear in the "Old World" is to be seen in Mies van der Rohe's "Glas-Hochhaus" project (1919) in Berlin and – as a built example – in the Fiat-Lingotto production facilities in

Turin by the engineer Giacomo Mattè Trucco (1916–1920). Le Corbusier, in his "Vers une Architecture" (1922) made reference to this and to the American feats of engineering as exemplary models for modern European architects almost ten years after Walter Gropius addressed the issue of America's importance for European modernism in a lecture on "Monumentale Kunst und Industriebau"[81] and in an article on "Die Entwicklung Moderner Industriebaukunst"[82]. And it was Adolf Loos, again 15 years before Gropius, who in the Viennese press tried to caustically explain the advantages of American Modernity to a turn-of-the-century metropolitan public in order to create breathing space for the creative potential to grow in that city in the Beaux-Arts tradition.

The "America Project" of early European modernism also provoked a further challenge by the modern definition of architectonic space disseminated by the Wasmuth in his publications on the work of Frank Lloyd Wright, from which later the new term of space-time simultaneousness or "transparency" was derived.

[81] Volkwang–Museum in Hagen, 1911.

[82] Yearbook of the Deutsche Werkbund of 1913.

Sources and References to Further Reading

Badinter, E. and R., Condorcet. Un intellectuel en politique, Paris 1988.

Bancroft, H.H., The Book of the Fair. An Historical and Descriptive Presentation of the World's Columbian Exposition at Chicago in 1893, New York 1894.

Banham, R., A Concrete Atlantis. U.S. Industrial Building and European Modern Architecture 1900–1925, The MIT Press, Cambridge, Mass. and London 1986.

Boyle, B.M., Architectural Practice in America, 1865–1965, in: S. Kostof (Ed.), The Architect. Chapters in the History of the Profession, New York 1977.

Brown, D.J., Bridges – Three Thousand Years of Defying Nature, London 1993.

Calvert, M.A., The Mechanical Engineer in America, 1830–1910, Baltimore, Maryland, 1967.

Condit, C.W., The Chicago School of Architecture. A History of Commercial and Public Building in the Chicago Area 1875–1925, Chicago/London 1964 (first edition 1952).

Condit, C.W., American Building. Materials and Techniques from the Beginning of the Colonial Settlements to the Present, Chicago/London 1982 (first edition 1968).

Donovan, A., Education, industry, and the American university. A transatlantic perspective, in: R. Fox and A. Guagnini (Ed.), Education, technology and industrial performance in Europe, 1850–1939, Cambridge 1993.

Draper, J., The Ecole des Beaux-Arts and the Architectural Profession in the United States: The Case of John Galen Howard, in: S. Kostof (Ed.), The Architect. Chapters in the History of the Profession, New York/Oxford 1977.

Emmerson, G.S., Engineering Education: A Social History, Newton Abbot, Devon and New York 1973.

Finch, J.K., A History of the School of Engineering. Columbia University, New York 1954.

Finch, J.K., The Story of Engineering, Garden City, New York, 1960.

Freitag, J.K., Architectural Engineering. New York/London 1904.

Giedion, S., Space, Time and Architecture. The Growth of a New Tradition, Cambridge, Mass., 1941.

The Great Exhibition of 1851, exhib. cat. (Ed. C.H. Gibbs-Smith), London 1950 (Reprint 1964).

Greene, J.C., American Science in the Age of Jefferson, Ames, Iowa, 1984.

Hounshell, D.A., From the American System to Mass Production 1800–1932, Baltomore and London 1985 (first edition 1984).

Pevsner, N., A History of Building Types, London 1976.

Pevsner, N, H. Honour, J. Fleming (Ed.), Lexikon der Weltarchitektur, München 1992.

Roth, L.M., A Concise History of American Architecture, New York 1979.

Rowe, C., Chicago Frame, in: 'Architectural Design', December 1970.

Semper, G., Der Stil in den technischen und tektonischen Künsten oder praktische Ästhetik. Ein Handbuch für Techniker, Künstler und Kunstfreunde, Vol. 2 (1863), Mittenwald 1977 (New ed.).

Steinman, D.B., The Builders of the Bridge. The Story of John Roebling and His Son, New York 1945.

Syrett, H.C. (Ed.), American Historical Documents, New York 1960.

Weigold, Marilyn E., Silent Builder. Emily Warren Roebling and the Brooklyn Bridge, Port Washington, NY., New York City and London 1984.

Wickenden, W.E., A Comparative Study of Engineering Education in the United States and in Europe, in: Bulletin N° 16 of the Investigation of Engineering Education (Ed. The Society for the Promotion of Engineering Education), Lancaster, Pennsylvania, 1929.

Woods, Mary N., From Craft to Profession. The Practice of Architecture in Nineteenth-Century America, Berkeley, 1999.

The Origins of Scientific and Technical Education in Great Britain

Henry Cole (1808–1882), Initiator and Organizer of the Great Exhibition of 1851 in London. (Author's archive)

As in France, Germany, Switzerland, Italy and America the initiative to establish scientific and technical institutions of education for architects and engineers in Great Britain also corresponded to an event of important national policy: the decisive moment here is represented by the first World Exhibition in London's Hyde Park of 1851.

Great Britain's economic predominance established by its early industrialization since the 1750s was, one hundred years later, threatened by important industrial developments particularly in France and America. The exhibition organizers, especially Henry Cole and Lyon Playfair, as well as the interested public were shocked by the modernity of certain products exhibited by these two countries at the "Great Exhibition of the Works of Industry of All Nations" in 1851.

Commissioned by Prince Albert to visit the "Exposition de l'Industrie Nationale" in Paris in 1849, Henry Cole returned to London and on 30 June 1849 submitted to the Royal Society of Arts a proposal to stage an international exhibition (the very first one to encompass a universal character) in London.[1] What is remarkable is that Cole endeavored to expand or change what was as yet a national project into one that was international and to obtain Prince Albert's

[1] Cole, 1884, p.116–205.

National or international exhibition? A facsimile of handwritten corrections carried out by Prince Albert on 30 June 1849. (Cole, 1884, Vol. I, after p.126)

for this site could be made to the Crown.
It was a question whether this exhibition should be exclusively limited to British Industry – It was considered that, whilst it appears an error to fix any limitation to the productions of Machinery, Science & Taste which are of no country but belong as a whole to the civilized world, particular advantage to British Industry might be derived from placing it in fair competition with that of other Nations. —
It was agreed that such Industrial Machinery and

support. A handwritten report documents the decision that was evidently made after just a short period of discussion.[2]

[2] Cole, 1884, p.123ff.

These circumstances are important in the context of further developments in that the "fair competition with that of other Nations" intended by the exhibition organizers confronted Great Britain with the state of the art of its competitors. On display at the Great Exhibition, for example, was the American MacCormick company's newest reaper, which attracted a great deal of attention: "In agriculture, it appears that the machine will be as important as the spinning-jenney and power-loom in manufactures."[3]

[3] Gibbs-Smith, 1950, p.82 (legend to fig.70)

American reaping machine by McCormick. I.L.N. July 1851

MacCormick Reaper, Exhibit at the Crystal Palace of 1851. (Author's archive)

Its construction of interchangeable parts combined the aspects of manifold application, opportunities for repair and development, technical precision and optimal form, thus producing an industrial design of the highest quality.[4] French industrial achievements were also particularly impressive in that they exhibited a mature state of technical design in contrast to the traditional British exhibits.[5] As a consequence of this self-induced and most restorative of all shocks reverberating from the London Crystal

[4] Giedion, 1941 (part V, p.257ff.).

[5] Gray, 1937.

Palace competition, England was given its subsequent scientific and technical schools of higher learning.

Lyon Playfair (1818–1898), Spokesman for the "Lectures on the Exhibition". (Reid, 1899, p.ii)

"Lectures on the Results of the Exhibition of 1851"
Some of the organizers, including Henry Cole and particularly Lyon Playfair, Commissioner of the Society of Arts, attributed England's relative lack of industrial quality in comparison to the Continent to a deficiency in cooperation between science and industry. Particularly the French culture of scientific-technological and scientific-industrial higher education, which had an established tradition of maintaining a practice-oriented teaching program was now recognized after the Great Exhibition as an exemplary model: "In France they have the Ecole Polytechnique, Ecole des Mines, Ecole Centrale des Arts et Manufactures, and the Conservatoire des Arts et Metiers."[6]

[6] Hole, 1853, p.50.

In reference to the tradition of apprenticeship that over a long period of time allowed Great Britain to maintain a leading position among industrial nations, Playfair sought to combine the advantages it offered with those of a scientific culture of education: "We do not think that such schools can substitute a practical training in the workshops, the factory, or the office of the engineer; but we do think that a producer possessing a knowledge of natural forces will become a practical man in a shorter time than without it, and that he will know how to turn his practice to the best account. The promoters of industrial instruction do not, therefore, offer it as a substitute for practical training, but consider it to be a means by which the latter can be made more efficacious."[7]

[7] Playfair, cit. in: Hole, 1853, p.55f.; on Playfair, cf. also Reid, 1899.

John Locke and the System of Apprenticeship. This system was meant to augment and expand the British model of learning rather than to replace it. Britain's early industrial and technical accomplishments are after all to be traced back to this system of apprenticeship based on the philosophical and educational principles formulated by John Locke on the threshold to the industrial revolution.[8] In his major educational work "Some Thoughts concerning Education" (1693) he postulates the combination of handicraft, commercial or other workshops of practical activity with an individual scientific and humanist acqui-

[8] John Locke (1632–1704), philosopher, doctor, educator, political counselor and government official is considered to be the direct "ancestor" and

forerunner of the Enlightenment philosophy. The French "philosophes" of the "Siècle des Lumières" also based their work on his. Directly after Descartes he predominantly addressed questions of epistemology, which he published as "An Essay concerning Human Understanding" in 1689, just one year after the "Glorious Revolution" in England and one hundred years before the French Revolution.

sition of education. For more than 150 years this educational ideal provided the impetus for inventions at an exceptionally early point in time and produced industrial revolutions that had a formative influence on pioneers such as Thomas Telford or Joseph Paxton, just to mention two protagonists of building history. Since in the process of the industrial revolution the importance of knowledge and the theoretical reasoning of scientific findings and their application in social practice continued to increase rapidly, the English model lacked an institutional correlation between theory and praxis. Early French schools had on the other hand already developed operative methods such as the "industrial sciences". The École des Ponts et Chaussées founded in Paris in 1747 could already look back on an approximately hundred-year tradition at the time of the Great Exhibition! Britain was forced to catch up quickly.

As a consequence of the "lesson of 1851" a broad debate was launched and many calls were made on the part of scientific societies and institutions "to join education to practice" and to establish "systematic education", to create schools of design or even a central college of arts and manufactures, etc. The intellectual level of skilled workers was found to be unsatisfactory: "(...) the greater part of the workmen have not received any assistance whatever from books, or from any subject of study pursued out of the shop; indeed, it is notorious that some of our most skilled workmen are, in other respects, grossly ignorant, to the extent in one or two instances of not being able to read or write, experience and practice alone supplying the necessary skill."[9]

"The Journal of Design and Manufactures"

Critique and demands were not to be heard and made only after the Great Exhibition. In Henry Cole's Journal of Design questions concerning the conditions and quality of "Art manufactures" were a continuous issue of priority from the very first number published in March 1849.[10] And it is no coincidence that the considerable amount of 180,000 British Pounds[11] derived from the profits of the Great Exhibition initiated by Cole were to be applied to the creation of two important instruments that would establish the correlation

[9] Cit. after: Hole, 1853, p.161–168 (Appendices).

[10] In addition, Nicolette Gray points out further prominent critics of the Victorian approach to art, architecture and architectural education who were leading individuals of the Great Exhibition: Richard Redgrave (Juror to the 1851 Exhibition), Digby Wyatt (Secretary to the executive committee of the Exhibition), Owen Jones (Superintendent of the building operations in the Crystal Palace), and others.: Gray, 1937, p.49.

[11] Gibbs-Smith, 1950, p.34; Hobhouse, 1950, p.150–151.

between art and industry: the founding of a School of Design and of the Victoria and Albert Museum of applied art.[12] This was to complete the development that had already taken place in France, at first in particular according to the example provided by the Conservatoire National des Arts et Métiers. After the Great Exhibition Cole even published the École Centrale des Arts et Manufacture's complete program of instruction over three and a half pages in order to illustrate the special nature of the "industrial sciences".[13] At length he reported in detail on the activities of the French exhibition director Charles Dupin (graduate of the École Polytechnique and student of Durand) and on the prizes and medals awarded to French exhibitors.[14]

In his journal, which Giedion describes as a "fighting journal",[15] Cole developed the concept behind the newly planned Schools of Design and of the Mechanics' Institutions; he also published many letters on this issue. The Journal of Design became one of the most central platforms of discussion for the modern movement of "arts manufacture", a term that Cole formulated. In his journal opposite views were also published, including a detailed academically oriented suggestion to base the schools of design upon artistic principles[16]; this was juxtaposed by a pointed account written by Lyon Playfair: "(that) industrial training and intellectual development are synonymous terms."[17] All of these statements, however, did not go beyond the level of conception and were pragmatic in nature. A formal "curricular shape" as exhibited by every French school at its inception was lacking here during Britain's respective pioneering age. In this journal England's lack of "school culture" was also repeatedly pointed out as the reason for the call to establish a new program of education: "The Council (of the Society of Arts) believe that a radical cure for many imperfections of British manufactures will be found in a much more enlarged and liberal system of art-education than at present exists (...)."[18]

Interestingly enough, one of the first consequences of the Great Exhibition was to establish a School of Design for architecture, metalwork and handicraft design and to appoint Gottfried Semper as its director.[19]

[12] The ways in which the Exhibition surplus was to be put to use was hotly contested, as evident in many articles published in Cole's Journal of Design, e.g. in: Cole, 1852, p.48–53.

[13] Cole, 1852, p.118–121.

[14] Cole, 1852, p.32 and p.157f.

[15] Giedion, 1948 (part V.: Henry Cole's Journal of Design p.349).

[16] Dyce, Education of Artists and Designers, in: Cole, 1852, p.131–135 and p.165–167.

[17] Playfair, Industrial Education, in: Cole, 1852, p.141.

[18] Cole, 1852, p.52.

[19] Cf. a detailed account of Gottfried Semper above, also in: Giedion, 1948, p.357f.

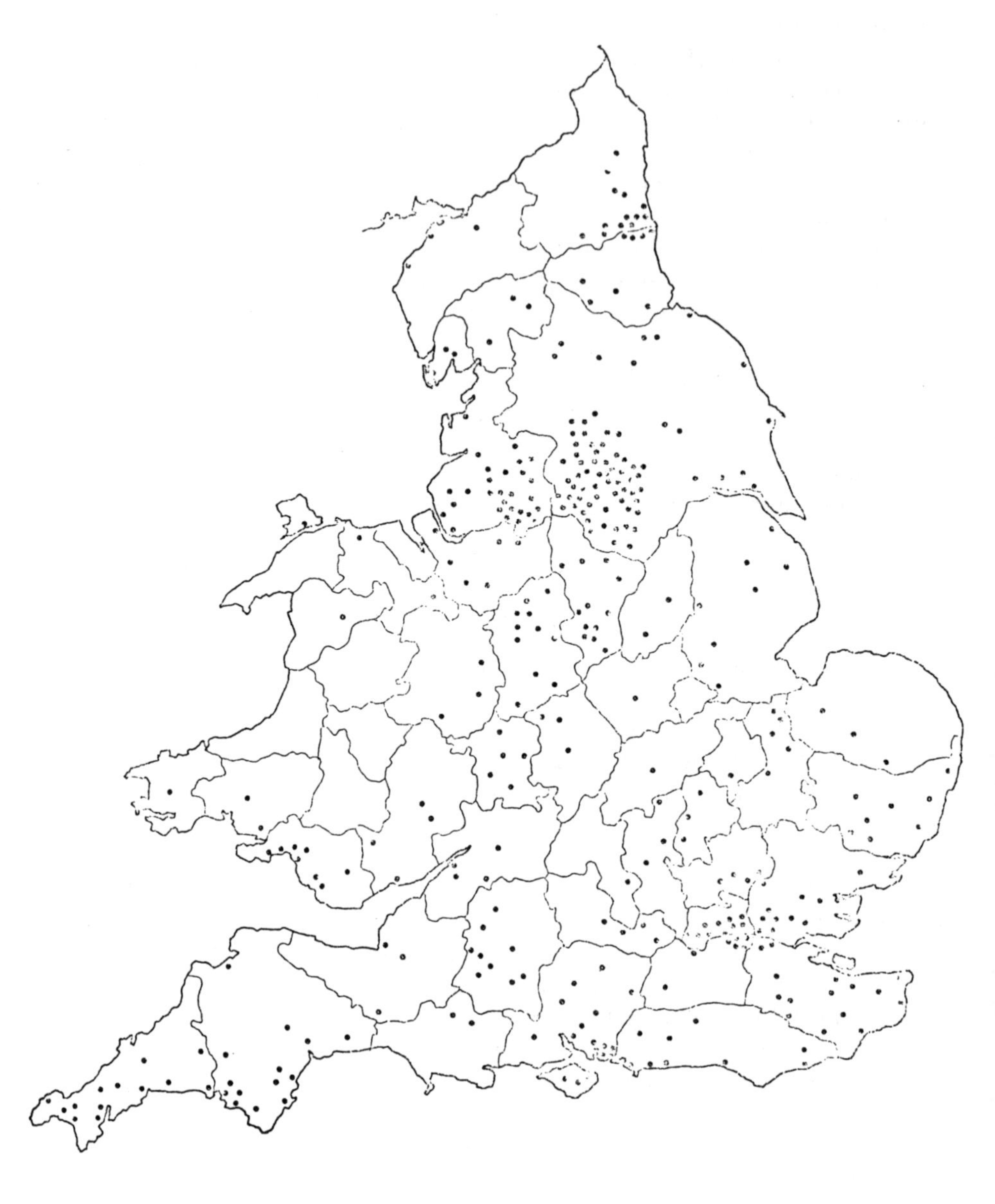

Geographic Distribution of the "Mechanics' Institutes" around 1851.
(Cardwell, 1972, p.75)

Mechanic Institutes

A second parallel movement initiated by the Society of Arts, Henry Cole and particularly by Lyon Playfair and others endeavored to establish a scientific and technical system of education: "In the establishment of institutions for industrial instruction, you, at the same time, create the wanting means for the advancements of science in this country."[20] About one year after the Great Exhibition, the 700 existing Mechanics' Institutes spread all across England (with 110,000 students), 220 were added to them with about 90,000 new students, all associated with the Society of Arts.[21] The promoters were scholars themselves such as Lyon Playfair (industrial chemist) or James Booth (mathematician). These institutes were to be transformed according to the French example of the Parisian École Centrale des Arts et Manufactures into industrial schools and central technical colleges, like the Birmingham and Midlands Institute of Industrial Education (opened 1853). It was within this context that the following generation of engineers was to be educated.[22]

The Society of Arts founded a Science and Art Department in 1853 which was to incorporate schools of design, museums, schools of mines and other scientific and technical institutes; its goal was to promote science, the skilled and technical arts and their industrial application.[23]

In addition, some of the scientific and technical schools established at this time were instituted at universities.[24] After further important steps had been taken additional opposition had to be overcome and a few more decades were to pass before a scientific-technical and scientific-industrial system of education developed, by the 1880s it became broadly established as an independent tradition. One of the milestones in this development is the National Conference that was called by the Society of Arts in 1868, which in turn set up an "acceleration committee", and the activation of a debate on the establishment of a Technical University in London (1877). The promoters of this movement, besides Henry Cole and Lyon Playfair, John Scott Russell, Bernhard Samuelson, Thomas H. Huxley and others, did not reach a

[20] Playfair, Lectures on the results of the Exhibition, 1852, cit. in: Cardwell, 1972, p.81.

[21] Cardwell, 1972, p.74f.

[22] Turner, 1927.

[23] Emmerson, 1973, p.170f.

[24] Cardwell, 1972, chap. 4, see also Turner, 1927.

breakthrough until 30 years after the Great Exhibition had taken place.[25]

Architects and engineers were also not able to reap the benefits of systematic and scientifically based technical and industrial instruction until the end of the 1870s.[26]

On the Education of Architects

The education of architects was also not characterized by school culture over a very long period of time: "In England (...) the only training available in the late eighteenth century was the emerging practice of articled pupilage in an architect's office, as traditionally introduced by Sir Robert Taylor. Additional training in drawing was otherwise gained in the company of painters and sculptors in various improvised art schools, such as the St. Martin's Lane Academy."[27] John Locke's model of learning also characterized the architect's training.[28] While the Society of Civil Engineers had already been founded in 1771, the Institute of British Architects was not established until 1834 (with the attribute Royal as of 1866) – the year of J.-N.-L Durand's death! It took a further 36 years until in 1870 the Royal Academy finally called a School of Architecture to life. It was directed by Phené Spiers, a student of the École des Beaux-Arts. As a result, an important part of architectural education was guided by the Beaux-Arts movement, similar to trends in America and at about the same time. And the "studio" model once again reinvigorated the system of patronage in England.[29]

Parallel to this, opposition to RIBA emerged as of 1842 in the form of a circle of young architects who established an Architectural Association (AA) and introduced a form of "self help curriculum" comprised of free discussions and critiques supplied by visiting lecturers. All initiatives were voluntary and only in 1891 was the AA willing to pay teachers and personnel.[30] In comparison to those schools that have been introduced in this book and that characterize the pioneering age of this modern profession, this institution has maintained its greater independent status up to the present date.

[25] Emmerson, 1973, chap.9.

[26] Wickenden, 1929, Artz, 1966, p.161.

[27] Wilton-Ely, in: Kostof, 1986, p.191.

[28] Jenkins delineates the history of education in architecture in England beginning with "Tudor Patronage". According to Jenkins, the institutionalized education of architecture begins with the establishment of King's College in London in 1840; its model of "Arts of Construction in connection with Civil Engineering and Architecture" is to be considered a methodological innovation: Jenkins, 1961 (esp. chap.8: Architectural Training in the Nineteenth Century, p.160–177).

[29] Fletcher, in: Gotch, 1934, p.85–98; on the academic approach, cf. also Briggs, 1927 (esp. chap. VIII).

[30] Wilton-Ely, in: Kostof, 1986, p.198.

Sources and References to Further Reading

Artz, F.B., The Development of Technical Education in France, Cambridge, Mass. and London 1966 (with extensive bibliography).
Briggs, M.S., The Architect in History, Oxford 1927 (esp. chapter VIII.).
Cardwell, D.S.L., The Organisation of Science in England, London 1972.
Cole, H. (Ed.), The Journal of Design and Manufactures: Vol. I. (March–August 1849), London 1849: Vol. VI. (September 1851–February 1852), London 1852.
Cole, H., Fifty Years of Public Work, London 1884 (Vol. I).
Dyce, W., Education of Artists and Designers, in: H. Cole (Ed.), 1852, p. 131–135 and p. 165–167.
Emmerson, G.S., Engineering Education. A Social History, New York 1973.
Fletcher, H.M., Architectural Education, in: J.A. Gotch (Ed.), The Growth and Work of the Royal Institute of British Architects 1834–1934, London 1934, p. 85–98.
Gibbs-Smith, C.H., The Great Exhibition of 1851, exhib. cat. on the occasion of its 100th year anniversary (Ed. Victoria & Albert Museum), London 1950 (reprinted 1964).
Giedion, S., Space, Time and Architecture. The Growth of a New Tradition, Cambridge, Mass., 1941.
Giedion, S., Mechanization takes Command, Oxford University Press, New York 1948.
Gray, N., Prophets of the Modern Movement, in: The Architectural Review, Vol. 81 (January–June 1937), London 1937, p. 49–50.
Hobhouse, Ch., 1851 and the Crystal Palace, London 1950.
Hole, J., An Essay on the History and Management of Literary, Scientific & Mechanics' Institutions, London 1853 (reprinted London 1970).
Jenkins, F., Architect and Patron. A survey of professional relations and practice in England from the sixteenth century to the present day, London 1961 (esp. chapter 8: Architectural Training in the Nineteenth Century).
Playfair, L., Lectures on the Results of the Exhibition (Ed.: Society of Arts, 2 Vols., London 1852).
Playfair, L., Industrial Education, in: H. Cole (Ed.), 1852, p. 140–144 and p. 163f.
Playfair, L., Lectures on Industrial Education on the Continent, cit. in: J. Hole, 1853, p. 55f.
Reid, W. (Ed.), Memoirs and Correspondence of Lyon Playfair, London 1899 (reprinted Jemimaville, Scotland 1976).
Turner, D.M., History of Science Teaching in England (esp. chapter V.), London 1927.
Wickenden, W.E., A Comparative Study of Engineering Education in the United States and in Europe, in: Bulletin N° 16 of the Investigation of Engineering Education (Ed.: The Society for the Promotion of Engineering Education), Lancaster, Pa. 1929.
Wilton-Ely, J., The Rise of the Professional Architect in England, in: S. Kostof (Ed.), The Architect. Chapters in the History of the Profession, New York and Oxford 1986 (first edition 1977), p. 180–208.

Outlook – Perspectives on a Development

In 1794 scholars dedicated to the philosophy of the Enlightenment founded the École Polytechnique in Paris; in 1806 the old art academy, now called the École des Beaux-Arts, was reopened by Napoleon. In his book "Space, Time, Architecture", Giedion considers these "two poles" that had faced each other since the beginning of the century as the point of departure for two epochal issues: "Along what lines should the training of an architect proceed? What is the relation of the engineer to the architect? What special functions are proper to each?"[1]

[1] Giedion, 1941, p.147.

The development of a new tradition. Within the framework of a polytechnical model of instruction Durand developed an architectural theory which included the didactic and systematic interconnection of environmental references, spatial qualities, of rules related to material construction and building structures, underpinned by a systematized method of design that brought about the modern architect. As a consequence, the goal of Durand's architecture course and the school as a whole was "to establish a connection between science and life, to bring out the practical applications to industry of discoveries in the mathematical and physical sciences.", as Giedion remarks.[2] The "Méthode Durand" was able to pave the way for generations of architects (and engineers) to exert a formative influence on the emerging industrial era.

[2] Giedion, 1941, p.147.

The École Polytechnique's model of instruction in the first two decades of the 19th century served as an example or reference point to founders of many other higher technical schools of learning throughout Europe and the U.S.. Some of Durand's students established the École Centrale des Arts et Manufactures in Paris in 1829, thus launching the second stage of modern architectural and engineering education by its institution of an industrially oriented model of learning and teaching, which again motivated many other new schools or at least the modernization of already existing schools such as Karlsruhe.

Both models of education initially conceived of the architect and engineer as a unity, particularly because their profession was based on the same scientific, technological and humanistic background. The cooperation between architects and engineers, "Polytechniciens", "Centraux" and even Beaux-Arts graduates was a natural occurrence in the 19th century as is exemplified in the construction of the Palais des Machines on the occasion of the Paris Exposition of 1889.

However, around the middle of the century the course of study was divided as a result of the creation of specific professional departments at schools of technical higher learning. Leading the way was Karlsruhe, then Zurich. The "Eidgenössische Polytechnikum" in Zurich was the first school in the world to establish a number of complete and autonomous courses of study for various modern professional fields after 1855. Nevertheless, the school's founders recognized the need to provide architects and engineers with a common curriculum and they thus institutionalized a system of preparatory schools in the natural sciences and humanities, including the establishment of a department of general educational courses directly within the "Poly", which offered courses, for example, in literature, languages, history, politics, economy, jurisprudence and later also philosophy. For this department renowned teachers were appointed, signalizing the great importance the school placed on the humanistic responsibility carried by the architect and engineer in a modern world.

Transitions and Constants. The history of the modern model of instruction for architects and engineers is a history of transitions: pioneers also always set the course for their successors who would assess, perfect, augment and seek broader application of their inventions.

Durand's teaching and learning model was carried by the Enlightenment conviction that men are generally capable of learning and acquiring education. Durand created a system of learning steps and strategies that allowed young students to apply themselves to building assignments in a systematic and methodical manner, thereby attaining a solution that was also rationally based.

In contrast to this is the principle of learning by model or by imitation within the context of master studios as maintained by the École des Beaux-Arts.[3] While here the work on building assignments was connected to the "Maître's" personal and professional style and training took place according to a catalog of solution types organized in increasing order – from the simple house to urban planning – Durand's system of instruction was a dynamic model: since it was conceived of as a method from the very beginning it also could establish itself independently of the teaching personality. The teacher became a conveyor – not only of his own built examples. The principle of methodic procedure in class instruction facilitated the mastering of all possible tasks and as such became available to all interested and well-prepared students.

[3] Walter Gropius' Bauhaus-pedagogy was also formed by this tradition, though it was industrially oriented.

On the "Basic Design Course (Grundkurs)" at the ETH Zurich's Department of Architecture 1961–1996. An example of the continuous development of instruction in architecture (based on Durand and with consideration to industrial innovations provided by Reynaud and Mary), and also of the shift in paradigm with respect to the approach to architecture and to the model of instruction is the "Basic Design Course" taught by the Department of Architecture at the ETH in Zurich. It was worked out and established by Bernhard Hoesli, Heinz Ronner and Hans Ess in the beginning of the 1960s and further reinforced by a viewpoint of cultural and architectural history through Adolf Max Vogt and Paul Hofer (urban planning) and continued by Herbert E. Kramel until 1996.[4]

[4] On Hoesli's background, cf. Caragonne, 1995, and Jansen et al., 1989; on Ronner, cf. Pfammatter, 1991; Kramel, 1983; as an overview on design education in Europe after 1945, cf. Pfammatter, 1998.

The "Basic Design Course" takes up the principle of easing the learning process by establishing a methodical instrumentarium that allows for a systematic approach to assignments and to solution strategies. With it the "Basic Design Course" endeavors to overcome the doctrinaire "modern constriction" that emerged between the wars and especially the "new Bauhaus" influence in architectural education. In addition, confronted by the "eclectic situation" as a consequence of the 1960s it seeks to reactivate the elementary principles of the "Métier": "It will be demonstrated that although the problem is a different one, the method of procedure is not. In order to illustrate the interaction between use, construction and space the relationships between singular problems will be presented in the most elementary manner possible, e.g. by the example of the skeleton – skin issue. As a result, the consequences for designing structural systems are also illustrated."[5]

[5] Hoesli, 1965.

Kramel's succeeding "Basic Design Course" takes up the direction inspired by Hoesli and develops further the methodological model of instruction: "The basic design course 85 and the model provided by Hoesli seeks to convey the basics of further study. It is not an introduction. – It is based on steps in teaching and learning, the goals and methods of which are open to verification. – It considers the learning process as a working and forming process. – It utilizes exercise as the smallest teaching unit (and not the project). – It takes up the ideas of typical problems ("Problem-typen"). – It conveys knowledge and creates experience primarily by means of the teaching program (and not primarily through the teacher)."[6]

[6] Kramel, 1986 and 1997.

The "quantum leap" instigated by Durand's architectural theory at the École Polytechnique was made possible by the introduction of "utility" and "economic structures" as generators of design; Mary's architecture course taught the spatial and form-determining power of structure, building technique and material. The ETH's "Basic Design Course" ultimately addressed the systematic design process itself as one of the most basic experiences of learning. As a result every model of teaching and learning was based on the other and assumed the former's characteristics to be a matter of course.

Reference to practice in the "Basic Design Course" is established by choosing tasks and exercises from which the conditions of the building program and site, the spatial, structural and formal problems become evident and recognized by means of that method as "problem types" and thus are systematically solved; "problem types", no longer "building types" or "solution types" become the landmarks of the learning process. In contrast to the original 19th century models the increasing complexity of reality can now be more extensively taken into consideration in building assignments. From this a broader field of interpretation emerges that challenges the students' ability to argue independently and formulate their positions precisely.

PRIMÄRSYSTEM UND DECKENSYSTEM

siehe auch KONTEXT DECKE/DACH

Vom tragwerkstechnischen Standpunkt lässt sich die Zugehörigkeit bestimmter Decken zu bestimmten Bauweisen, die sich in einem Primärsystem manifestieren, feststellen. Setzen wir die Regel, dass das Auflagerangebot eines Primärsystems für Deckenlasten nie ungenutzt bleiben soll, kann eine solche Zuordnung eindeutig erfolgen. Es ist z.B. eindeutig, dass bei Verwendung von gerichteten Deckensystemen (z.B. einfache Balkenlage, Tonnengewölbe, Hourdisdecken, Durisol/Lecca/Stahlton-Deckenelemente etc.) eine Schottenbauweise entstehen möchte bzw. keine Massivbauweise nötig ist bzw. ohne die Hilfsmassnahmen von Unterzügen kein Skelettbau möglich ist. Umgekehrt ist offensichtlich, dass eine Schottenbauweise die Verwendung der genannten Deckensysteme bevorzugt. Eine Massivbauweise hingegen bietet Auflager für eine allseitig aufliegende Decke.

AUFLAGERSYSTEMATIK UEBER EINEM FELD

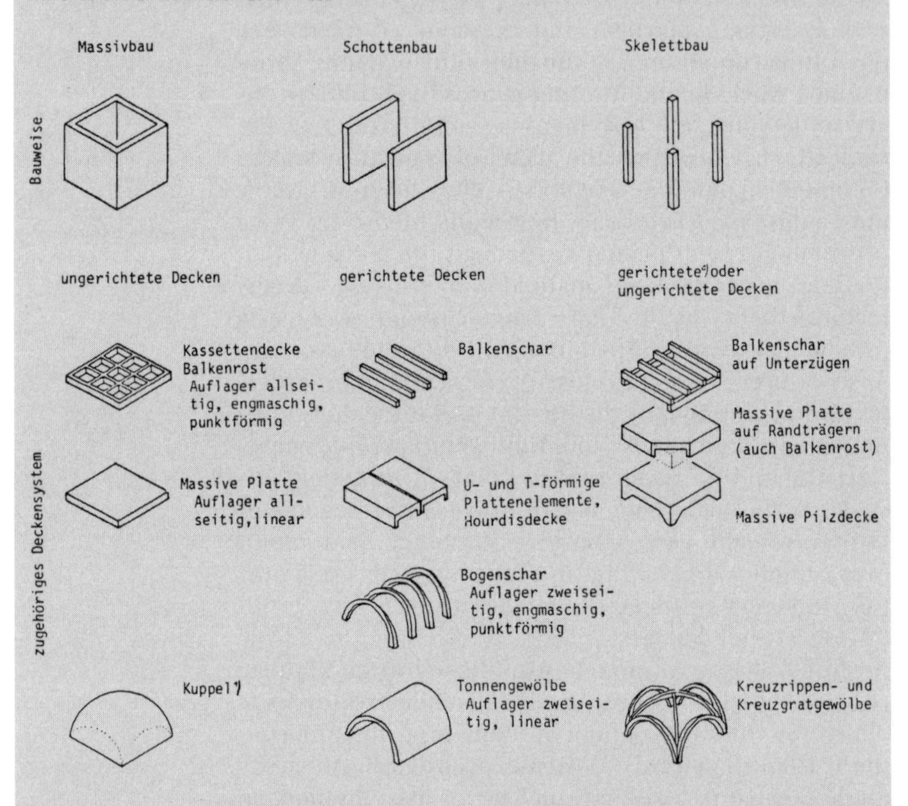

¹) Die Verbindung von Kuppel und Quadrat führt zum Stutz-, Pendentiv- oder Trompengewölbe.

²) Ein gerichtetes System bedingt die Einführung von Unterzügen. Meist ist in diesem Fall auch die Stützenstellung gerichtet.

Diese Gesetzmässigkeiten gelten auch bei den nicht orthogonalen Varianten der Grundtypen. Ein Massivbau kann z.B. kreisrund sein, Schotten können nichtparallel stehen und die Stützenstellung eines Skelettbaus kann unregelmässig sein.

Die Zuordnungstendenz bestimmter Deckensysteme zu bestimmten Bauweisen ist evident. Sollte bei einem Massivbau aus Gründen des Nichtvorhandenseins oder der Nichtanwendbarkeit oder aus Gründen der nachträglichen leichteren Perforierbarkeit der Decke eine Balkendecke verwendet werden, "drückt" die Gesetzmässigkeit der Zuordnung von Deckensystem zu Primärsystem bei der Oeffnungsbildung "durch": Eine Wand, welche keine Balkenauflager enthält lässt andere, z.B. breitere Oeffnungen zu, als eine Auflagerwand, wo Oeffnungen schmal zu werden tendieren, um die gleichmässige Abtragung der Lasten in der Wandfläche nicht zu stören.

Using H. Ronner's teaching material "Baustruktur" in constructive design (1988) a clear parallel to Durand's architecture course of the year 1821[7], which shows the correlation between spatial, structural and formal problems, is worthy of being mentioned in conclusion; the comparison illustrates the power of the original polytechnical model of instruction in combination with its later industrial orientation.

opposite page: "Building structure": the methodical correlation between spatial, structural and formal decisions in structural design (teaching material "Kontext" at the ETH by H. Ronner in 1988). (Ronner, 1988, p.13)

What is revealing here is that Durand intuitively recognized and modern learning theory substantiated certain constants of "learning strategies" connected to the natural conditions of the human organism. These include, for example, the question of practical training: while imitating "good" solutions is hardly enough, "learning by doing" also does not automatically lead to independent thought and work either. Neither learning models nor praxis can do without "scores", "rules of language" or a "design repertoire". With the help of these tools experience is gained through exercise and the ability to transfer acquired experience is reinforced. The Zurich "Basic Design Course" once again marks the transition as a modern "school of thought" by its innovations of teaching and methodology. The author is at present working towards developing this model further on an interdisciplinary level at the University of Applied Sciences in Lucerne. This involves structural engineering, services and technical equipment engineering and industrial design. For this purpose new media are also being utilized.[8]

With their understanding of the "Métier" these three fundamental models established a tradition of renewal capable of regenerating the cultural humus of education and of broadening education, thus liberating it from the virtual shadow images Plato uses as a metaphor in his "cave parable" to describe intellectual bondage. In this respect, with his "Marche à suivre" Durand brought about a "Copernican revolution" in the history of architectural and engineering education.

[7] Cf. illus. "Éléments des Édifices" by Durand on p. 65.

[8] Cf. Pfammatter, 1999.

Sources and References to Further Reading

Caragonne, A., The Texas Rangers. Notes from an architectural underground, Cambridge, Mass. and London 1995.

Giedion, S., Space, Time and Architecture. The Growth of a New Tradition, Cambridge, Mass. 1941.

Jansen, J., Hu. Jörg, L. Maraini, Hp. Stöckli (Ed.), Architektur lehren. Bernhard Hoesli an der Architekturabteilung der ETH Zürich (Institute for History and Theory of Architecture at the ETH Zurich), Zurich 1989.

Hoesli, B., Ganzheitlich orientierte Architektenausbildung, in: Detail, N° 3, München 1965; Reprint in: G. Kueng, L. Maraini, H. Ronner (Ed.), Bernhard Hoesli. Architexte (lectures, articles, scripts, notes etc.), Zurich 1991.

Kramel, H.E., Der Konstruktionsunterricht an der Abteilung für Architektur der ETH Zurich, in: 'Schweizer Architekt' und Ingenieur, N° 47, Zurich 1983, p. 1111–1119.

Kramel, H.E., Die Lehre als Programm – on basic design (Institute for History and Theory of Architecture at the ETH Zürich), Zurich 1986.

Kramel, H.E., A Structural Approach to Basic Architectural Design. The Zurich Model 1985–1995, ETH Zurich, School of Architecture, Zurich 1997.

Pfammatter, U. (with H. Ronner), Heinz Ronner – Lehrtätigkeit an der ETH Zürich 1961–1991, 5 Vols., ETH Zurich 1991.

Pfammatter, U. (with H.E. Kramel), Herbert E. Kramel – Protokolle. Zur Lehrtätigkeit an der ETH Zürich 1961–1996, ETH Zurich (in preparation).

Pfammatter, U., Ausbilden nach 1945. Zur Geschichte einer didaktischen Entwicklung im Kontext der modernen Architekturbewegung – eine Skizze, in: 'Schweizer Architekt und Ingenieur', N° 37, Zurich 1998, p. 679–682.

Pfammatter, U., An Interdisciplinary Design Education Model for Architecture and Building Engineering. The "I.D.E.M.-Case Study" at the University of Applied Sciences/Lucerne School of Engineering and Architecture (ICED 99 Munich), University of Technology Munich 1999 (Conference Vol. 2, p. 1271–1274).

Ronner, H., Baustruktur (lecture script 'Kontext', N° 78), Zurich 1988.

Index